TOPICS IN
PROBABILITY

TOPICS IN
PROBABILITY

Narahari Prabhu
Cornell University, USA

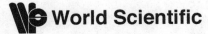
World Scientific

NEW JERSEY · LONDON · SINGAPORE · BEIJING · SHANGHAI · HONG KONG · TAIPEI · CHENNAI

Published by

World Scientific Publishing Co. Pte. Ltd.

5 Toh Tuck Link, Singapore 596224

USA office: 27 Warren Street, Suite 401-402, Hackensack, NJ 07601

UK office: 57 Shelton Street, Covent Garden, London WC2H 9HE

British Library Cataloguing-in-Publication Data
A catalogue record for this book is available from the British Library.

TOPICS IN PROBABILITY

ISBN-13 978-981-4335-47-8
ISBN-10 981-4335-47-9

Typeset by Stallion Press
Email: enquiries@stallionpress.com

Printed in Singapore.

Now I've understood
Time's magic play:

Beating his drum he rolls out the show,
Shows different images
And then gathers them in again

Kabir (1450–1518)

CONTENTS

PREFACE

In this monograph we treat some topics that have been of some importance and interest in probability theory. These include, in particular, analytic characteristic functions, the moment problem, infinitely divisible and self-decomposable distributions.

We begin with a review of the measure-theoretical foundations of probability distributions (Chapter 1) and characteristic functions (Chapter 2).

In many important special cases the domain of characteristic functions can be extended to a strip surrounding the imaginary axis of the complex plane, leading to analytic characteristic functions. It turns out that distributions that have analytic characteristic functions are uniquely determined by their moments. This is the essence of the moment problem. The pioneering work in this area is due to C. C. Heyde. This is treated in Chapter 3.

Infinitely divisible distributions are investigated in Chapter 4. The final Chapter 5 is concerned with self-decomposable distributions and triangular arrays. The coverage of these topics as given by Feller in his 1971 book is comparatively modern (as opposed to classical) but is still somewhat diffused. We give a more compact treatment.

N. U. Prabhu
Ithaca, New York
January 2010

ABBREVIATIONS

Term	Abbreviation
characteristic function	c.f.
distribution function	d.f
if and only if	iff
Laplace transform	L.T.
probability generating function	p.g.f
random variable	r.v.

Terminology: We write $x \stackrel{d}{=} y$ if the r.v.'s x, y have the same distribution.

Chapter 1

Probability Distributions

1.1. Elementary Properties

A function F on the real line is called a probability distribution function if it satisfies the following conditions:

(i) F is non-decreasing: $F(x + h) \geq F(x)$ for $h > 0$;
(ii) F is right-continuous: $F(x+) = F(x)$;
(iii) $F(-\infty) = 0$, $F(\infty) \leq 1$.

We shall say that F is *proper* if $F(\infty) = 1$, and F is *defective* otherwise.

Every probability distribution induces an assignment of probabilities to all Borel sets on the real line, thus yielding a probability measure P. In particular, for an interval $I = (a, b]$ we have $P\{I\} = F(b) - F(a)$. We shall use the same letter F both for the point function and the corresponding set function, and write $F\{I\}$ instead of $P\{I\}$. In particular

$$F(x) = F\{(-\infty, x]\}.$$

We shall refer to F as a probability distribution, or simply a *distribution*.

A point x is an *atom* if it carries positive probability (weight). It is a *point of increase* iff $F\{I\} > 0$ for every open interval I containing x.

1

A distribution F is *concentrated* on the set A if $F(A^c) = 0$, where A^c is the complement of A. It is *atomic* if it is concentrated on the set of its atoms. A distribution without atoms is *continuous*.

As a special case of the atomic distribution we have the *arithmetic* distribution which is concentrated on the set $\{k\lambda(k = 0, \pm 1, \pm 2, \ldots)\}$ for some $\lambda > 0$. The largest λ with this property is called the *span* of F.

A distribution is *singular* if it is concentrated on a set of Lebesgue measure zero. Theorem 1.1 (below) shows that an atomic distribution is singular, but there exist singular distributions which are continuous.

A distribution F is *absolutely continuous* if there exists a function f such that

$$F(A) = \int_A f(x)\,dx.$$

If there exists a second function g with the above property, then it is clear that $f = g$ almost everywhere, that is, except possibly on a set of Lebesgue measure zero. We have $F'(x) = f(x)$ almost everywhere; f is called the *density* of F.

Theorem 1.1. *A probability distribution has at most countably many atoms.*

Proof. Suppose F has n atoms $x_1, x_2, \ldots, x_n$ in $I = (a, b]$ with $a < x_1 < x_2 < \cdots < x_n \le b$ and weights $p(x_k) = F\{x_k\}$. Then

$$\sum_{k=1}^{n} p(x_k) \le F\{I\}.$$

This shows that the number of atoms with weights $> \frac{1}{n}$ is at most equal to n. Let

$$D_n = \{x : p(x) > 1/n\};$$

then the set D_n has at most n points. Therefore the set $D = \cup D_n$ is at most countable.   □

Theorem 1.2 (Jordan decomposition). *A probability distribution F can be represented in the form*

$$F = pF_a + qF_c \tag{1.1}$$

where $p \geq 0, q \geq 0, p + q = 1, F_a, F_c$ are both distributions, F_a being atomic and F_c continuous.

Proof. Let $\{x_n, n \geq 1\}$ be the atoms and $p = \sum p(x_n), q = 1 - p$. If $p = 0$ or if $p = 1$, the theorem is trivially true. Let us assume that $0 < p < 1$ and for $-\infty < x < \infty$ define the two functions

$$F_a(x) = \frac{1}{p} \sum_{x_n \leq x} p(x_n), \quad F_c(x) = \frac{1}{q}[F(x) - pF_a(x)]. \tag{1.2}$$

Here F_a is a distribution because it satisfies the conditions (i)–(iii) above. For F_c we find that for $h > 0$

$$q[F_c(x+h) - F_c(x)] = F(x+h) - F(x) - \sum_{x < x_n \leq x+h} p(x_n) \geq 0, \tag{1.3}$$

which shows that F_c is non-decreasing. Letting $h \to 0$ in (1.3) we find that

$$q[F_c(x_+) - F_c(x)] = F(x_+) - F(x) = 0,$$

so that F is right-continuous. Finally, $F_c(-\infty) = 0$, while $F_c(\infty) = 1$. Therefore F_c is a distribution. □

Theorem 1.3 (Lebesgue decomposition). *A probability distribution can be represented as the sum*

$$F = pF_a + qF_{sc} + rF_{ac} \tag{1.4}$$

where $p \geq 0, q \geq 0, r \geq 0, p + q + r = 1$, F_a is an atomic distribution, F_{sc} is a continuous but singular distribution and F_{ac} is an absolutely continuous distribution.

Proof.   By the Lebesgue decomposition theorem on measures we can express F as

$$F = aF_s + bF_{ac}, \tag{1.5}$$

where $a \geq 0, b \geq 0, a + b = 1, F_s$ is a singular distribution and F_{ac} is an absolutely continuous distribution. Applying Theorem 1.2 to F_s we find that $F_s = p_1 F_a + q_1 F_{sc}$, where $p_1 \geq 0, q_1 \geq 0, p_1 + q_1 = 1$. Writing $p = ap_1, q = aq_1, r = b$ we arrive at the desired result (1.4). $\square$

Remark.   Although it is possible to study distribution functions and measures without reference to random variables (r.v.) as we have done above, it is convenient to start with the definition

$$F(x) = P\{X \leq x\}$$

where X is a random variable defined on an appropriate sample space.

1.2. Convolutions

Let F_1, F_2 be distributions and F be defined by

$$F(x) = \int_{-\infty}^{\infty} F_1(x - y)dF_2(y) \tag{1.6}$$

where the integral obviously exists. We call F the convolution of F_1 and F_2 and write $F = F_1 * F_2$. Clearly $F_1 * F_2 = F_2 * F_1$.

Theorem 1.4.   *The function F is a distribution.*

Proof.   For $h > 0$ we have

$$F(x + h) - F(x) = \int_{-\infty}^{\infty} [F_1(x - y + h) - F_1(x - y)]dF_2(y) \geq 0 \tag{1.7}$$

so that F is non-decreasing. As $h \to 0$,

$$F_1(x - y + h) - F_1(x - y) \to F_1(x - y+) - F_1(x - y) = 0;$$

since

$$|F_1(x - y + h) - F_1(x - y)| \le 2, \quad \int_{-\infty}^{\infty} 2dF_2(y) = 2,$$

the right side of (1.7) tends to 0 by the dominated convergence theorem. Therefore $F(x_+) - F(x) = 0$, so that F is right-continuous. Since $F_1(\infty) = 1$ the dominated convergence theorem gives $F(\infty) = 1$. Similarly $F(-\infty) = 0$. Therefore F is a distribution. $\qquad \square$

Theorem 1.5. *If F_1 is continuous, so is F. If F_1 is absolutely continuous, so is F.*

Proof. We have seen in Theorem 1.4 that the right-continuity of F_1 implies the right-continuity of F. Similarly the left-continuity of F_1 implies that of F. It follows that if F_1 is continuous, so is F.

Next let F_1 be absolutely continuous, so there exists a function f_1 such that

$$F_1(x) = \int_{-\infty}^{x} f_1(u)du.$$

Then

$$F(x) = \int_{-\infty}^{\infty} dF_2(y) \int_{-\infty}^{x} f_1(u - y)du$$

$$= \int_{-\infty}^{x} \left\{ \int_{-\infty}^{\infty} f_1(u - y)dF_2(y) \right\} du$$

so that F is absolutely continuous, with density

$$f(x) = \int_{-\infty}^{\infty} f_1(x - y)dF_2(y). \qquad (1.8)$$

$\qquad \square$

Remarks.
1. If X_1, X_2 are independent random variables with distributions F_1, F_2, then the convolution $F = F_1 * F_2$ is the distribution of their

sum $X_1 + X_2$. For

$$F(z) = P\{X_1 + X_2 \le z\} = \iint_{x+y\le z} dF_1(x)dF_2(y)$$

$$= \int_{-\infty}^{\infty} dF_2(y) \int_{-\infty}^{z-y} dF_1(x) = \int_{-\infty}^{\infty} F_1(z-y)dF_2(y).$$

However, it should be noted that *dependent* random variables X_1, X_2 may have the property that the distribution of their sum is given by the convolution of their distributions.

2. The converse of Theorem 1.5 is false. In fact two singular distributions may have a convolution which is absolutely continuous.

3. The *conjugate* of any distribution F is defined as the distribution $\tilde{F}$, where

$$\tilde{F}(x) = 1 - F(-x_-).$$

If F is the distribution of the random variable X, then $\tilde{F}$ is the distribution of $-X$. The distribution F is *symmetric* if $F = \tilde{F}$.

4. Given any distribution F, we can *symmetrize* it by defining the distribution $^\circ F$, where

$$^\circ F = F * \tilde{F}.$$

It is seen that $^\circ F$ is a symmetric distribution. It is the distribution of the difference $X_1 - X_2$, where X_1, X_2 are independent variables with the same distribution F.

1.3. Moments

The moment of order $\alpha > 0$ of a distribution F is defined by

$$\mu_\alpha = \int_{-\infty}^{\infty} x^\alpha dF(x)$$

provided that the integral converges absolutely, that is,

$$\nu_\alpha = \int_{-\infty}^{\infty} |x|^\alpha dF(x) < \infty;$$

ν_α is called the absolute moment of order α. Let $0 < \alpha < \beta$. Then for $|x| \le 1$ we have $|x|^\alpha \le 1$, while for $|x| > 1$ we have $|x|^\alpha \le |x|^\beta$.

Thus we can write $|x|^\alpha \le |x|^\beta + 1$ for all x and so

$$\int_{-\infty}^{\infty} |x|^\alpha dF(x) \le \int_{-\infty}^{\infty} (1 + |x|^\beta)dF(x) = 1 + \int_{-\infty}^{\infty} |x|^\beta dF(x).$$

This shows that the existence of the moment of order β implies the existence of all moments of order $\alpha < \beta$.

Theorem 1.6. *The moment μ_α of a distribution F exists iff*

$$x^{\alpha-1}[1 - F(x) + F(-x)] \tag{1.9}$$

is integrable over $(0, \infty)$.

Proof. For $t > 0$ an integration by parts yields the relation

$$\int_{-t}^{t} |x|^\alpha dF(x) = -t^\alpha [1 - F(t) + F(-t)]$$

$$+ \alpha \int_{0}^{t} x^{\alpha-1}[1 - F(x) + F(-x)]dx. \tag{1.10}$$

From this we find that

$$\int_{-t}^{t} |x|^\alpha dF(x) \le \alpha \int_{0}^{t} x^{\alpha-1}[1 - F(x) + F(-x)]dx$$

so that if (1.9) is integrable over $(0, \infty)$, ν_α (and therefore μ_α) exists. Conversely, if ν_α exists, then since

$$\int_{|x|>t} |x|^\alpha dF(x) > |t|^\alpha [1 - F(t) + F(-t)]$$

the first term on the right side of (1.10) vanishes as $t \to \infty$ and the integral there converges as $t \to \infty$. □

Theorem 1.7. *Let*

$$\nu(t) = \int_{-\infty}^{\infty} |x|^t dF(x) < \infty$$

for t in some interval I. Then $\log \nu(t)$ is a convex function of $t \in I$.

Proof.   Let $a \geq 0, b \geq 0, a + b = 1$. Then for two functions ψ_1, ψ_2 we have the Hölder inequality

$$\int_{-\infty}^{\infty} |\psi_1(x)\psi_2(x)| dF(x) \leq \left[\int_{-\infty}^{\infty} |\psi_1(x)|^{1/a} dF(x)\right]^a$$

$$\times \left[\int_{-\infty}^{\infty} |\psi_2(x)|^{1/b} dF(x)\right]^b$$

provided that the integrals exist. In this put $\psi_1(x) = x^{at_1}, \psi_2(x) = x^{bt_2}$, where $t_1, t_2 \in I$. Then

$$\nu(at_1 + bt_2) \leq \nu(t_1)^a \nu(t_2)^b \tag{1.11}$$

or taking logarithms,

$$\log \nu(at_1 + bt_2) \leq a \log \nu(t_1) + b \log \nu(t_2)$$

which establishes the convexity property of $\log \nu$. $\square$

Corollary 1.1 (Lyapunov's inequality).   *Under the hypothesis of Theorem 1.7, $\nu_t^{\frac{1}{t}}$ is non-decreasing for $t \in I$.*

Proof.   Let $\alpha, \beta \in I$ and choose $a = \alpha/\beta$, $t_1 = \beta$, $b = 1 - a$, $t_2 = 0$. Then (1.11) reduces to

$$\nu_\alpha \leq \nu_\beta^{\alpha/\beta} (\alpha \leq \beta)$$

where we have written $\nu_t = \nu(t)$. $\square$

1.4. Convergence Properties

We say that I is an interval of continuity of a distribution F if I is open and its end points are not atoms of F. The whole line $(-\infty, \infty)$ is considered to be an interval of continuity.

    Let $\{F_n, n \geq 1\}$ be a sequence of proper distributions. We say that the sequence converges to F if

$$F_n\{I\} \to F\{I\} \tag{1.12}$$

for every bounded interval of continuity of F. If (1.12) holds for every (bounded or unbounded) interval of continuity of F, then the convergence is said to be *proper*, and otherwise *improper*. Proper convergence implies in particular that $F(\infty) = 1$.

Examples

1. Let F_n be uniform in $(-n, n)$. Then for every bounded interval contained in $(-n, n)$ we have

$$F_n\{I\} = \int_I \frac{dx}{2n} = \frac{|I|}{2n} \to 0 \quad \text{as } n \to \infty$$

where $|I|$ is the length of I. This shows that the convergence is improper.

2. Let F_n be concentrated on $\{\frac{1}{n}, n\}$ with weight $1/2$ at each atom. Then for every bounded interval I we have

$$F_n\{I\} \to 0 \quad \text{or} \quad 1/2$$

according as I does not or does contain the origin. Therefore the limit F is such that it has an atom at the origin, with weight $1/2$. Clearly F is not a proper distribution.

3. Let F_n be the convolution of a proper distribution F with the normal distribution with mean zero and variance n^{-2}. Thus

$$F_n(x) = \int_{-\infty}^{\infty} F(x - y) \frac{n}{\sqrt{2\pi}} e^{-(1/2)n^2 y^2} \, dy$$

$$= \int_{-\infty}^{\infty} F(x - y/n) \frac{1}{\sqrt{2\pi}} e^{-(1/2)y^2} \, dy.$$

For finite a, b we have

$$\int_a^b dF_n(x) = \int_{-\infty}^{\infty} [F(b - y/n) - F(a - y/n)] \frac{1}{\sqrt{2\pi}} e^{-(1/2)y^2} \, dy$$

$$\to F(b_-) - F(a_-) \quad \text{as } n \to \infty$$

by the dominated convergence theorem. If a, b are points of continuity of we can write

$$F_n\{(a, b)\} \to F\{(a, b)\} \tag{1.13}$$

so that the sequence $\{F_n\}$ converges properly to F.

If X is a random variable with the distribution F and Y_n is an independent variable with the above normal distribution, then we know that F_n is the distribution of the sum $X + Y_n$. As $n \to \infty$, it is obvious that the distribution of this sum converges to that of X. This justifies the definition of convergence which requires (1.13) to hold only for points of continuity a, b.

Theorem 1.8 (Selection theorem). *Every sequence $\{F_n\}$ of distributions contains a subsequence $\{F_{n_k}, k \geq 1\}$ which converges (properly or improperly) to a limit F.*

Theorem 1.9. *A sequence $\{F_n\}$ of proper distributions converges to F iff*

$$\int_{-\infty}^{\infty} u(x) dF_n(x) \to \int_{-\infty}^{\infty} u(x) dF(x) \qquad (1.14)$$

for every function u which is bounded, continuous and vanishing at $\pm\infty$. If the convergence is proper, then (1.14) holds for every bounded continuous function u.

The proofs of these two theorems are omitted.

Chapter 2

Characteristic Functions

2.1. Regularity Properties

Let F be a probability distribution. Then its characteristic function (c.f.) is defined by

$$\phi(\omega) = \int_{-\infty}^{\infty} e^{i\omega x} dF(x) \tag{2.1}$$

where $i = \sqrt{-1}, \omega$ real. This integral exists, since

$$\int_{-\infty}^{\infty} |e^{i\omega x}| dF(x) = \int_{-\infty}^{\infty} dF(x) = 1. \tag{2.2}$$

Theorem 2.1. *A c.f. ϕ has the following properties:*

(a) $\phi(0) = 1$ *and* $|\phi(\omega)| \leq 1$ *for all* ω.
(b) $\phi(-\omega) = \bar{\phi}(\omega)$, *and* $\bar{\phi}$ *is also a c.f.*
(c) Re ϕ *is also a c.f.*

Proof.   (a) We have

$$\phi(0) = \int_{-\infty}^{\infty} dF(x) = 1, \quad |\phi(\omega)| \leq \int_{-\infty}^{\infty} |e^{i\omega x}| dF(x) = 1.$$

(b) $\bar{\phi}(\omega) = \int_{-\infty}^{\infty} e^{-i\omega x} F(dx) = \phi(-\omega)$. Moreover, let $\tilde{F}(x) = 1 - F(-x_-)$. Then

$$\int_{-\infty}^{\infty} e^{i\omega x} \tilde{F}\{dx\} = \int_{-\infty}^{\infty} e^{-i\omega x} \tilde{F}\{-dx\} = \int_{-\infty}^{\infty} e^{-i\omega x} F\{dx\}.$$

Thus $\phi(-\omega)$ is the c.f. of $\tilde{F}$, which is a distribution.

(c) Re $\phi = \frac{1}{2}\phi + \frac{1}{2}\bar{\phi} = $ c.f. of $\frac{1}{2}F + \frac{1}{2}\tilde{F}$, which is a distribution. □

Theorem 2.2. *If ϕ_1, ϕ_2 are c.f.'s, so is their product $\phi_1\phi_2$.*

Proof.   Let ϕ_1, ϕ_2 be the c.f.'s of F_1, F_2 respectively and consider the convolution

$$F(x) = \int_{-\infty}^{\infty} F_1(x - y) dF_2(y).$$

We know that F is a distribution. Its c.f. is given by

$$\phi(\omega) = \int_{-\infty}^{\infty} e^{i\omega x} dF(x) = \int_{-\infty}^{\infty} e^{i\omega x} \int_{-\infty}^{\infty} dF_1(x - y) dF_2(y)$$

$$= \int_{-\infty}^{\infty} e^{i\omega y} dF_2(y) \int_{-\infty}^{\infty} e^{i\omega(x-y)} dF_1(x - y)$$

$$= \phi_1(\omega)\phi_2(\omega).$$

Thus the product $\phi_1\phi_2$ is the c.f. of the convolution $F_1 * F_2$. □

Corollary 2.1. *If ϕ is a c.f., so is $|\phi|^2$.*

Proof.   We can write $|\phi|^2 = \phi\bar{\phi}$, where $\bar{\phi}$ is a c.f. by Theorem 2.1(b). □

Theorem 2.3. *A distribution F is arithmetic iff there exists a real $\omega_0 \neq 0$ such that $\phi(\omega_0) = 1$.*

Proof.   (i) Suppose that the distribution is concentrated on $\{k\lambda, \lambda > 0, k = 0, \pm 1, \pm 2, \ldots\}$ with the weight p_k at $k\lambda$. Then the c.f. is given by

$$\phi(\omega) = \sum_{-\infty}^{\infty} p_k e^{i\omega k\lambda}.$$

Clearly $\phi(2\pi/\lambda) = 1$.

(ii) Conversely, let $\phi(\omega_0) = 1$ for $\omega_0 \neq 0$. This gives

$$\int_{-\infty}^{\infty} (1 - e^{i\omega_0 x}) dF(x) = 0.$$

Therefore

$$\int_{-\infty}^{\infty} (1 - \cos \omega_0 x) dF(x) = 0$$

which shows that the points of increase of F are among $\frac{2k\pi}{\omega_0} (k = 0, \pm 1, \pm 2, \ldots)$. Thus the distribution is arithmetic. $\qquad\square$

Corollary 2.2. *If $\phi(\omega) = 1$ for all ω, then the distribution is concentrated at the origin.*

Remarks.
1. If F is the distribution of a random variable, then we can write

$$\phi(\omega) = E(e^{i\omega X})$$

so that the c.f. is the expected value of $e^{i\omega X}$. We have $\phi(-\omega) = E(e^{-i\omega X})$, so that $\phi(-\omega)$ is the c.f. of the random variable $-X$. This is Theorem 2.1(b).
2. If X_1, X_2 are two independent random variables with c.f.'s ϕ_1, ϕ_2, then

$$\phi_1(\omega)\phi_2(\omega) = E[e^{i\omega(X_1 + X_2)}]$$

so that the product $\phi_1\phi_2$ is the c.f. of the sum $X_1 + X_2$. This is only a special case of Theorem 2.2, since the convolution $F_1 * F_2$ is not necessarily defined for independent random variables.
3. If ϕ is the c.f. of the random variable X, then $|\phi|^2$ is the c.f. of the symmetrized variable $X_1 - X_2$, where X_1, X_2 are independent variables with the same distribution as X.

Theorem 2.4. (a) *ϕ is uniformly continuous.*

(b) *If the n-th moment exists, then the n-th derivative exists and is a continuous function given by*

$$\phi^{(n)}(\omega) = \int_{-\infty}^{\infty} e^{i\omega x}(ix)^n dF(x). \tag{2.3}$$

(c) *If the n-th moment exists, then ϕ admits the expansion*

$$\phi(\omega) = 1 + \sum_1^n \mu_n \frac{(i\omega)^n}{n!} + 0(\omega^n) \ (\omega \to 0). \qquad (2.4)$$

Proof.   (a) We have

$$\phi(\omega + h) - \phi(\omega) = \int_{-\infty}^{\infty} e^{i\omega x}(e^{ihx} - 1)dF(x) \qquad (2.5)$$

so that

$$|\phi(\omega + h) - \phi(\omega)| \leq \int_{-\infty}^{\infty} |e^{ihx} - 1|dF(x)$$

$$\leq 2\int_{-\infty}^{\infty} |\sin(hx/2)|dF(x).$$

Now

$$\int_{x<-A, x>B} |\sin(hx/2)|dF(x) \leq \int_{x<-A, x>B} dF(x) < \varepsilon$$

by taking A, B large, while

$$\int_{-A}^{B} |\sin(hx/2)|dF(x) \leq \eta \int_{-A}^{B} dF(x) < \eta.$$

since $|\sin(hx/2)| < \eta$ for h small. Therefore $|\phi(\omega+h)-\phi(\omega)| \to 0$ as $h \to 0$, which proves uniform continuity.

(b) We shall prove (2.3) for $n = 1$, the proof being similar for $n > 1$. We can write (2.5) as

$$\frac{\phi(\omega + h) - \phi(\omega)}{h} = \int_{-\infty}^{\infty} e^{i\omega x} \cdot \frac{e^{ihx} - 1}{h}dF(x). \qquad (2.5')$$

Here

$$\left| e^{i\omega x} \cdot \frac{e^{ihx} - 1}{h} \right| \leq \left| \frac{e^{ihx} - 1}{h} \right| \leq |x|$$

and

$$\int_{-\infty}^{\infty} |x|dF(x) < \infty$$

by hypothesis. Moreover $(e^{ihx} - 1)/h \to ix$ as $h \to 0$. Therefore letting $h \to 0$ in (2.5') we obtain by the dominated convergence theorem that

$$\frac{\phi(\omega + h) - \phi(\omega)}{h} \to \int_{-\infty}^{\infty} ixe^{i\omega x}dF(x)$$

as required. Clearly, this limit is continuous.

(c) We have

$$e^{i\omega x} = \sum_{n=0}^{n} \frac{(i\omega x)^n}{n!} + o(\omega^n x^n) \quad (\omega \to 0)$$

so that

$$\int_{-\infty}^{\infty} e^{i\omega x}dF(x) = 1 + \sum_{n=1}^{n} \frac{(i\omega)^n}{n!}\mu_n + \int_{-\infty}^{\infty} o(\omega^n x^n)dF(x),$$

where the last term on the right side is seen to be $o(\omega^n)$. □

Remark. The converse of (b) is not always true: thus $\phi'(\omega)$ may exist, but the mean may not. A partial converse is the following:

Suppose that $\phi^{(n)}(\omega)$ exists. If n is even, then the first n moments exist, while if n is odd, the first $n-1$ moments exist.

2.2. Uniqueness and Inversion

Theorem 2.5 (uniqueness). *Distinct distributions have distinct c.f.'s.*

Proof. Let F have the c.f. ϕ, so that

$$\phi(\omega) = \int_{-\infty}^{\infty} e^{i\omega x}dF(x).$$

We have for $a > 0$

$$\int_{-\infty}^{\infty} \frac{a}{\sqrt{2\pi}} e^{-\frac{1}{2}a^2\omega^2 - i\omega y}\phi(\omega)d\omega$$

$$= \int_{-\infty}^{\infty} \frac{a}{\sqrt{2\pi}} e^{-\frac{1}{2}a^2\omega^2 - i\omega y} \int_{-\infty}^{\infty} e^{i\omega x}dF(x)$$

$$= \int_{-\infty}^{\infty} dF(x) \int_{-\infty}^{\infty} e^{i\omega(x-y)} \frac{a}{\sqrt{2\pi}} e^{-\frac{1}{2}a^2\omega^2}d\omega,$$

the inversion of integrals being clearly justified. The last integral is the c.f. (evaluated at $x-y$) of the normal distribution with mean 0 and variance a^{-2}, and therefore equals $e^{-(x-y)^2/2a^2}$. We therefore obtain the identity

$$\frac{1}{2\pi}\int_{-\infty}^{\infty} e^{-\frac{1}{2}a^2\omega^2 - i\omega y}\phi(\omega)d\omega = \int_{-\infty}^{\infty} \frac{1}{\sqrt{2\pi}a}e^{-\frac{1}{2a^2}(y-x)^2}dF(x)$$

(2.6)

for all $a > 0$. We note that the right side of (2.6) is the density of the convolution $F * N_a$, where N_a is the normal distribution with mean 0 and variance a^2. Now if G is a second distribution with the c.f. ϕ, it follows from (2.6) that $F * N_a = G * N_a$. Letting $a \to 0_+$ we find that $F \equiv G$ as required. ☐

Theorem 2.6 (inversion). (a) *If the distribution F has c.f. ϕ and $|\phi(\omega)/\omega|$ is integrable, then for $h > 0$*

$$F(x+h) - F(x) = \frac{1}{2\pi}\int_{-\infty}^{\infty} e^{-i\omega x} \cdot \frac{1 - e^{-i\omega h}}{i\omega}\phi(\omega)d\omega. \quad (2.7)$$

(b) *If $|\phi|$ is integrable, then F has a bounded continuous density f given by*

$$f(x) = \frac{1}{2\pi}\int_{-\infty}^{\infty} e^{-i\omega x}\phi(\omega)d\omega. \quad (2.8)$$

Proof.   (b) From (2.6) we find that the density f_a of $F_a = F * N_a$ is given by

$$f_a(x) = \frac{1}{2\pi}\int_{-\infty}^{\infty} e^{-\frac{1}{2}a^2\omega^2 - i\omega x}\phi(\omega)d\omega. \quad (2.9)$$

Here the integrand is bounded by $|\phi(\omega)|$, which is integrable by hypothesis. Moreover, as $a \to 0_+$, the integrand $\to e^{-i\omega x}\phi(\omega)$. Therefore by the dominated convergence theorem as $a \to 0_+$,

$$f_a(x) \to \frac{1}{2\pi}\int_{-\infty}^{\infty} e^{-i\omega x}\phi(\omega)d\omega = f(x) \quad \text{(say)}.$$

Clearly, f is bounded and continuous. Now for every bounded interval I we have

$$F_a\{I\} = \int_I f_a(x)dx.$$

Letting $a \to 0_+$ in this we obtain

$$F\{I\} = \int_I f(x)dx$$

if I is an interval of continuity of F. This shows that f is the density of F, as required.

(a) Consider the uniform distribution with density

$$u_h(x) = \frac{1}{h} \quad \text{for } -h < x < 0, \text{ and } = 0 \text{ elsewhere.}$$

Its convolution with F has the density

$$f_h(x) = \int_{-\infty}^{\infty} u_h(x-y)dF(y) = \int_x^{x+h} \frac{1}{h}dF(y) = \frac{F(x+h) - F(x)}{h}$$

and c.f.

$$\phi_h(\omega) = \phi(\omega) \cdot \int_{-\infty}^{\infty} e^{i\omega x} u_h(x)dx = \phi(\omega) \cdot \frac{1 - e^{-i\omega h}}{i\omega h}.$$

By (b) we therefore obtain

$$\frac{F(x+h) - F(x)}{h} = \frac{1}{2\pi} \int_{-\infty}^{\infty} e^{-i\omega x} \phi(\omega) \cdot \frac{1 - e^{-i\omega h}}{i\omega h} d\omega$$

provided that $|\phi(\omega)(1 - e^{-i\omega h})/i\omega|$ is integrable. This condition reduces to condition that $|\phi(\omega)/\omega|$ is integrable. $\qquad \square$

2.3. Convergence Properties

Theorem 2.7 (continuity theorem). *A sequence $\{F_n\}$ of distributions converges properly to a distribution F iff the sequence $\{\phi_n\}$ of their c.f.'s converges to ϕ, which is continuous at the origin. In this case ϕ is the c.f. of F.*

Proof.   (i) If $\{F_n\}$ converges properly to F, then

$$\int_{-\infty}^{\infty} u(x)dF_n(x) \rightarrow \int_{-\infty}^{\infty} u(x)dF(x)$$

for every continuous and bounded function u. For $u(x) = e^{i\omega x}$ it follows that $\phi_n(\omega) \rightarrow \phi(\omega)$ where ϕ is the c.f. of F. From Theorem 2.4(a) we know that ϕ is uniformly continuous.

(ii) Conversely suppose that $\phi_n(\omega) \rightarrow \phi(\omega)$, where ϕ is continuous at the origin. By the selection theorem there exists a subsequence $\{F_{n_k}, k \geq 1\}$ which converges to F, a possibly defective distribution. Using (2.6) we have

$$\frac{a}{\sqrt{2\pi}} \int_{-\infty}^{\infty} e^{-i\omega y - \frac{1}{2}a^2\omega 2} \phi_{n_k}(\omega)d\omega = \int_{-\infty}^{\infty} e^{-\frac{1}{2a^2}(y-x)^2} dF_{n_k}(x).$$

Letting $k \rightarrow \infty$ in this we obtain

$$\frac{a}{\sqrt{2\pi}} \int_{-\infty}^{\infty} e^{-i\omega y - \frac{1}{2}a^2\omega 2} \phi(\omega)d\omega = \int_{-\infty}^{\infty} e^{-\frac{1}{2a^2}(y-x)^2} dF(x)$$

$$\leq F(\infty) - F(-\infty). \qquad (2.10)$$

Writing the first expression in (2.10) as

$$\frac{1}{\sqrt{2\pi}} \int_{-\infty}^{\infty} e^{-i\omega(y/a) - \frac{1}{2}\omega^2} \phi(\omega/a)d\omega \qquad (2.11)$$

and applying the dominated convergence theorem we find that (2.11) converges to $\phi(0) = 1$ as $a \rightarrow \infty$. By (2.10) it follows that $F(\infty) - F(-\infty) \geq 1$, which gives $F(-\infty) = 0$, $F(\infty) = 1$, so that F is proper. By (i) ϕ is the c.f. of F, and by the uniqueness theorem F is unique. Thus every subsequence $\{F_{n_k}\}$ converges to F. $\qquad \square$

Theorem 2.8 (weak law of large numbers). *Let $\{X_n, n \geq 1\}$ be a sequence of independent random variables with a common distribution and finite mean μ. Let $S_n = X_1 + X_2 + \cdots + X_n$ $(n \geq 1)$. Then as $n \rightarrow \infty$, $S_n/n \rightarrow \mu$ in probability.*

Proof. Let ϕ be the c.f. of X_n. The c.f. of S_n/n is then

$$E(e^{i\omega(S_n/n)}) = \phi(\omega/n)^n = [1 + i\mu(\omega/n) + 0(1/n)]^n \to e^{i\mu\omega}$$

as $n \to \infty$. Here $e^{i\mu\omega}$ is the c.f. of a distribution concentrated at the point μ. By the continuity theorem it follows that the distribution of S_n/n converges to this degenerate distribution. $\qquad\square$

Theorem 2.9 (central limit theorem). *Let* $\{X_n, n \geq 1\}$ *be a sequence of independent random variables with a common distribution and*

$$E(X_n) = \mu, \quad \text{Var}(X_n) = \sigma^2$$

(both being finite). Let $S_n = X_1 + X_2 + \cdots + X_n$ $(n \geq 1)$. *Then as* $n \to \infty$, *the distribution of* $(S_n - n\mu)/\sigma\sqrt{n}$ *converges to the standard normal.*

Proof. The random variables $(X_n - \mu)/\sigma$ have mean zero and variance unity. Let their common c.f. be ϕ. Then the c.f. of $(S_n - n\mu)/\sigma\sqrt{n}$ is

$$\phi(\omega/\sqrt{n})^n = [1 - \omega^2/2n + 0(1/n)]^n \to e^{-\frac{1}{2}\omega^2}$$

where the limit is the c.f. of the standard normal distribution. The desired result follows by the continuity theorem. $\qquad\square$

Remark. In Theorem 2.7 the convergence of $\phi_n \to \phi$ is uniform with respect to ω in $[-\Omega, \Omega]$.

2.3.1. *Convergence of types*

Two distributions F and G are said to be of the same type if

$$G(x) = F(ax + b) \tag{2.12}$$

with $a > 0$, b real.

Theorem 2.10. *If for a sequence $\{F_n\}$ of distributions we have*

$$F_n(\alpha_n x + \beta_n) \to G(x), \quad F_n(a_n x + b_n) \to H(x) \qquad (2.13)$$

for all points of continuity, with $\alpha_n > 0$, $a_n > 0$, and G and H are non-degenerate distributions, then

$$\frac{\alpha_n}{a_n} \to a, \quad \frac{\beta_n - b_n}{a_n} \to b \quad \text{and} \quad G(x) = H(ax+b) \qquad (2.14)$$

$(0 < a < \infty, |b| < \infty)$.

Proof. Let $H_n(x) = F_n(a_n x + b_n)$. Then we are given that $H_n(x) \to H(x)$ and also $H_n(\rho_n x + \sigma_n) = F_n(\alpha_n x + \beta_n) \to G(x)$, where

$$\rho_n = \frac{\alpha_n}{a_n}, \quad \sigma_n = \frac{\beta_n - b_n}{a_n}. \qquad (2.15)$$

With the obvious notations we are given that

$$\phi_n(\omega) \to \phi(\omega), \quad \psi_n(\omega) \equiv e^{-i\omega\sigma_n/\rho_n} \phi_n(\omega/\rho_n) \to \psi(\omega)$$

uniformly in $-\Omega \le \omega \le \Omega$. Let $\{\rho_{n_k}\}$ be a subsequence of $\{\rho_n\}$ such that $\rho_{n_k} \to a$ $(0 \le a \le \infty)$. Let $a = \infty$, then

$$|\psi(\omega)| = \lim |\psi_{n_k}(\omega)| = \lim |\phi_{n_k}(\omega/\rho_{n_k})| = |\phi(0)| = 1$$

uniformly in $[-\Omega, \Omega]$, so that ψ is degenerate, which is not true. If $a = 0$, then

$$|\phi(\omega)| = \lim |\phi_{n_k}(\omega)| = \lim |\psi_{n_k}(\rho_{n_k}\omega)| = |\psi(0)| = 1,$$

so that ϕ is degenerate, which is not true. So $0 < a < \infty$. Now

$$e^{-i\omega(\sigma_{n_k}/\rho_{n_k})} = \frac{\psi_{n_k}(\omega)}{\phi_{n_k}(\omega)} \to \frac{\psi(\omega)}{\phi(\omega)}$$

so that $\sigma_{n_k}/\rho_{n_k} \to a$ limit b/a (say). Also

$$\psi(\omega) = e^{-i\omega(b/a)} \phi(\omega/a). \qquad (2.16)$$

It remains to prove the uniqueness of the limit a. Suppose there are two subsequences of $\{\rho_n\}$ converging to a and a', and assume that

$a < a'$. Then the corresponding subsequences of $\{b_n\}$ converge to b, b' (say) From (2.16) we obtain

$$e^{-i\omega(b/a)}\phi(\omega/a) = e^{-i\omega(b'/a')}\phi(\omega/a')$$

and hence $|\phi(\omega/a)| = |\phi(\omega/a')|$ or

$$|\phi(\omega)| = |\phi(a/a')\omega| = |\phi(a^2/a'^2)\omega| = \cdots = |\phi(a^n/a'^n)\omega| = |\phi(0)| = 1.$$

This means that ϕ is degenerate, which is not true. So $a \not< a'$. Similarly $a \not> a'$. Therefore $a = a'$, as required. Since we have proved (2.16), the theorem is completely proved.   □

2.4. A Criterion for c.f.'s

A function f of a real variable ω is said to be non-negative definite in $(-\infty, \infty)$ if for all real numbers $\omega_1, \omega_2, \ldots, \omega_n$ and complex numbers $a_1, a_2, \ldots, a_n$

$$\sum_{r,s=1}^{n} f(\omega_r - \omega_s)a_r\bar{a}_s \geq 0. \tag{2.17}$$

For such a function the following properties hold.

(a) $f(0) \geq 0$. If in (2.17) we put $n = 2, \omega_1 = \omega, \omega_2 = 0, a_1 = a, a_2 = 1$ we obtain

$$f(0)(1 + |a|^2) + f(\omega)a + f(-\omega)\bar{a} \geq 0. \tag{2.18}$$

When $\omega = 0$ and $a = 1$ this reduces to $f(0) \geq 0$.

(b) $\bar{f}(\omega) = f(-\omega)$. We see from (2.18) that $f(\omega)a + f(-\omega)\bar{a}$ is real. This gives $\bar{f}(\omega) = f(-\omega)$.

(c) $|f(\omega)| \leq f(0)$. In (2.18) let us choose $a = \lambda\bar{f}(\omega)$ where λ is real. Then

$$f(0) + 2\lambda|f(\omega)|^2 + \lambda^2|f(\omega)|^2 f(0) \geq 0.$$

This is true for all λ, so $|f(\omega)|^4 \leq |f(\omega)|^2[f(0)]^2$ or $|f(\omega)| \leq f(0)$, as required.

Theorem 2.11. *A function ϕ of a real variable is the c.f. of a distribution iff it is continuous and non-negative definite.*

Proof.   (i) Suppose ϕ is a c.f.; that is,

$$\phi(\omega) = \int_{-\infty}^{\infty} e^{i\omega x} dF(x)$$

where F is a distribution. By Theorem 2.4(a), ϕ is continuous. Moreover,

$$\sum_{r,s=1}^{n} \phi(\omega_r - \omega_s) a_r \bar{a}_s$$

$$= \sum_{r,s=1}^{n} a_r \bar{a}_s \int_{-\infty}^{\infty} e^{i(\omega_r - \omega_s)x} dF(x)$$

$$= \int_{-\infty}^{\infty} \left(\sum_{1}^{n} a_r e^{i\omega_r n} \right) \left(\sum_{1}^{n} \bar{a}_s e^{-i\omega_s x} \right) dF(x)$$

$$= \int_{-\infty}^{\infty} \left| \sum_{i}^{n} a_r e^{i\omega_r x} \right|^2 dF(x) \geq 0$$

which shows that ϕ is non-negative definite.

(ii) Conversely, let ϕ be continuous and non-negative definite. Then considering the integral as the limit of a sum we find that

$$\int_{0}^{\tau} \int_{0}^{\tau} e^{-i(\omega - \omega')x} \phi(\omega - \omega') d\omega d\omega' \geq 0 \qquad (2.19)$$

for $\tau > 0$. Now consider

$$P_\tau(x) = \frac{1}{\tau} \int_{0}^{\tau} \int_{0}^{\tau} e^{-\lambda(\omega - \omega')x} \phi(\omega - \omega') d\omega d\omega'$$

$$= \int_{-\infty}^{\infty} e^{-isx} \phi_\tau(s) ds \qquad (2.20)$$

where

$$\phi_\tau(t) = \begin{cases} \left(1 - \frac{|t|}{\tau} \right) \phi(t) & \text{for } |t| \leq \tau \\ 0 & \text{for } |t| \geq \tau \end{cases}.$$

From (2.20) we obtain

$$\psi_\wedge(t) = \frac{1}{2\pi} \int_{-\wedge}^{\wedge} \left(1 - \frac{|\lambda|}{\wedge}\right) e^{it\lambda} P_\tau(\lambda) d\lambda$$

$$= \frac{1}{2\pi} \int_{-\infty}^{\infty} \phi_\tau(s) ds \int_{-\wedge}^{\wedge} \left(1 - \frac{|\lambda|}{\wedge}\right) e^{\lambda(t-s)\lambda} d\lambda$$

$$= \frac{1}{2\pi} \int_{-\infty}^{\infty} \frac{4\sin^2 \frac{1}{2} \wedge (s-t)}{\wedge(s-t)^2} \phi_\tau(s) ds \to \phi_\tau(t) \quad \text{as } \wedge \to \infty.$$

On the account of (2.19), ψ_λ is a c.f., and ϕ_τ is continuous at the origin. By the continuity theorem ϕ_τ is a c.f. Again

$$\phi_\tau(t) \to \phi(t) \quad \text{as } \tau \to \infty$$

and since ϕ is continuous at the origin it follows that ϕ is a c.f. as was to be proved.   □

Remark. This last result is essentially a theorem due to S. Bochner.

Remark on Theorem 2.7. If a sequence $\{F_n\}$ of distributions converges properly to a distribution F, then the sequence $\{\phi_n\}$ of their c.f.'s converges to ϕ, which is the c.f. of F and the convergence is uniform in every finite interval.

Proof.   Let $A < 0$, $B > 0$ be points of continuity of F. We have

$$\phi_n(\omega) - \phi(\omega) = \int_{-\infty}^{\infty} e^{i\omega x} F_n\{dx\} - \int_{-\infty}^{\infty} e^{i\omega x} F\{dx\}$$

$$= \int_{x<A, x>B} e^{i\omega x} F_n\{dx\} - \int_{x<A, x>B} e^{i\omega x} F\{dx\}$$

$$+ \left[\int_A^B e^{i\omega x} F_n\{dx\} - \int_A^B e^{i\omega x} F\{dx\} \right]$$

$$= I_1 + I_2 + I_3 \quad \text{(say)}.$$

We have

$$I_3 = \int_A^B e^{i\omega x} F_n\{dx\} - \int_A^B e^{i\omega x} F\{dx\}$$

$$= \{e^{i\omega x}[F_n(x) - F(x)]\}_A^B - i\omega \int_A^B e^{i\omega x}[F_n(x) - F(x)]dx$$

and so

$$|I_3| = |F_n(B) - F(B)| + |F_n(A) - F(A)|$$
$$+ |\omega| \int_A^B |F_n(x) - F(x)| dx.$$

Given $\varepsilon > 0$ we can make

$$|F_n(B) - F(B)| < \varepsilon/9, \quad |F_n(A) - F(A)| < \varepsilon/9$$

for n sufficiently large. Also, since $|F_n(x) - F(x)| \leq 2$ and $F_n(x) \to F(x)$ at points of continuity of F, we have for $|\omega| < \Omega$

$$|\omega| \int_A^B |F_n(x) - F(x)| dx \leq \Omega \int_A^B |F_n(x) - F(x)| dx < \varepsilon/9.$$

Thus

$$|I_3| < \varepsilon/3.$$

Also for A, B sufficiently large

$$|I_1| \leq \left| \int_{x<A, x>B} e^{i\omega x} F_n\{dx\} \right| \leq 1 - F_n(B) + F_n(A) < \frac{1}{3}\varepsilon$$

$$|I_2| \leq \left| \int_{x<A, x>B} e^{i\omega x} F_n\{dx\} \right| \leq 1 - F_n(B) - F_n(A) < \frac{1}{3}\varepsilon.$$

The results follow from the last three inequalities.   □

2.5. Problems for Solution

1. Consider the family of distributions with densities $f_a(-1 \leq a \leq 1)$ given by

$$f_a(x) = f(x)[1 + a \sin(2\pi \log x)]$$

where $f(x)$ is the log-normal density

$$f(x) = \frac{1}{\sqrt{2\pi}} x^{-1} e^{-1/2(\log x)^2} \quad \text{for } x > 0.$$

$$= 0 \quad \text{for } x \leq 0.$$

Show that f_a has exactly the same moments as f. (Thus the log-normal distribution is not uniquely determined by its moments).

2. Let $\{p_k, k \geq 0\}$ be a probability distribution, and $\{F_n, n \geq 0\}$ a sequence of distributions. Show that

$$\sum_{n=0}^{\infty} p_n F_n(x)$$

is also a distribution.

3. Show that $\phi(\omega) = e^{\lambda(e^{-|\omega|}-1)}$ is a c.f., and find the corresponding density.

4. A distribution is concentrated on $\{\pm 2, \pm 3, \ldots\}$ with weights

$$p_k = \frac{c}{k^2 \log |k|} (k = \pm 2, \pm 3, \ldots)$$

where c is such that the distribution is proper. Find its c.f. ϕ and show that ϕ' exists but the mean does not.

5. Show that the function $\phi(\omega) = e^{-|\omega|^\alpha} (\alpha > 2)$ is not a c.f.

6. If a c.f. ϕ is such that $\phi(\omega)^2 = \phi(c\omega)$ for some constant c, and the variance is finite, show that ϕ is the c.f. of the normal distribution.

7. A degenerate c.f. ϕ is factorized in the form $\phi = \phi_1\phi_2$, where ϕ_1 and ϕ_2 are c.f.'s. Show that ϕ_1 and ϕ_2 are both degenerate.

8. If the sequence of c.f.'s $\{\phi_n\}$ converges to a c.f. ϕ and $\omega_n \to \omega_0$, show that $\phi_n(\omega_n) \to \phi_n(\omega_0)$.

9. If $\{\phi_n\}$ is a sequence of c.f.'s such that $\phi_n(\omega) \to 1$ for $-\delta < \omega < \delta$, then $\phi_n(\omega) \to 1$ for all ω.

10. A sequence of distributions $\{F_n\}$ converges properly to a non-degenerate distribution F. Prove that the sequence $\{F_n(a_n x + b_n)\}$ converges to a distribution degenerate at the origin iff $a_n \to \infty$ and $b_n = 0(a_n)$.

Chapter 3

Analytic Characteristic Functions

3.1. Definition and Properties

Let F be a probability distribution and consider the transform

$$\phi(\theta) = \int_{-\infty}^{\infty} e^{\theta x} dF(x) \qquad (3.1)$$

for $\theta = \sigma + i\omega$, where σ, ω are real and $i = \sqrt{-1}$. This certainly exists for $\theta = i\omega$. Since

$$\left| \int_A^B e^{\theta x} dF(x) \right| \leq \int_A^B e^{\sigma x} dF(x), \qquad (3.2)$$

$\phi(\theta)$ exists if $\int_{-\infty}^{\infty} e^{\sigma x} dF(x)$ is finite. Clearly, the integrals

$$\int_0^{\infty} e^{\sigma x} dF(x), \quad \int_{-\infty}^0 e^{\sigma x} dF(x) \qquad (3.3)$$

converge for $\sigma < 0$, $\sigma > 0$ respectively. Suppose there exist numbers α, β $(0 < \alpha, \beta \leq \infty)$ such that the first integral in (3.3) converges for $\sigma < \beta$ and the second for $\sigma > -\alpha$, then

$$\int_{-\infty}^{\infty} e^{\sigma x} dF(x) < \infty \quad \text{for } -\alpha < \sigma < \beta. \qquad (3.4)$$

In this case $\phi(\theta)$ converges in the strip $-\alpha < \sigma < \beta$ of the complex plane, and we say (in view of Theorem 3.1 below) that F has an analytic c.f. ϕ. If $\alpha = \beta = \infty$ the c.f. is said to be entire (analytic on the whole complex plane).

27

The following examples show that a distribution need not have an analytic c.f. and also that there are distributions with entire c.f.'s. The conditions under which an analytic c.f. exists are stated in Theorem 3.5.

Examples

Distribution	c.f.	Regions of existence
Binomial: $f(n,k)' = \binom{n}{k} p^k q^{n-k}$	$(q + pe^\theta)^n$	whole plane
Normal: $f(x) = \dfrac{1}{\sqrt{2\pi}} e^{-\frac{1}{2}x^2}$	$e^{\frac{1}{2}\theta^2}$	whole plane
Cauchy: $f(x) = \dfrac{1}{\pi} \cdot \dfrac{1}{1+x^2}$	$e^{-|\theta|}$	$\sigma = 0$
Gamma: $f(x) = e^{-\lambda x} \lambda^\alpha \dfrac{x^{\alpha-1}}{\Gamma(\alpha)}$	$\left(1 - \dfrac{\theta}{\lambda}\right)^{-\alpha}$	$\sigma < \lambda$
Laplace: $f(x) = \dfrac{1}{2} e^{-|x|}$	$(1 - \theta^2)^{-1}$	$-1 < \sigma < 1$
Poisson: $f(k) = e^{-\lambda} \dfrac{\lambda^k}{k!}$	$e^{\lambda(e^\theta - 1)}$	whole plane

Theorem 3.1. *The c.f. ϕ is analytic in the interior of the strip of its convergence.*

Proof.   Let
$$I = \frac{\phi(\theta+h) - \phi(\theta)}{h} - \int_{-\infty}^{\infty} xe^{\theta x} dF(x)$$
where the integral converges in the interior of the strip of convergence, since for $\delta > 0$,
$$\left| \int_{-\infty}^{\infty} xe^{\theta x} dF(x) \right| \le \int_{-\infty}^{\infty} |x| e^{\sigma x} dF(x) \le \int_{-\infty}^{\infty} e^{\delta|x| + \sigma x} dF(x)$$
and the last integral is finite for $-\alpha + \delta < \sigma < \beta - \delta$. We have
$$I = \int_{-\infty}^{\infty} e^{\theta x} \left(\frac{e^{hx} - 1 - hx}{h} \right) dF(x)$$
$$= \int_{-\infty}^{\infty} e^{\theta x} (h(x^2/2!) + h^2 x^3/3! + \cdots) dF(x).$$

Therefore

$$|I| \leq \int_{-\infty}^{\infty} e^{\sigma x} |h| |x|^2 (1 + |hx|/1! + |hx|^2/2! + \cdots) dF(x)$$

$$\leq |h| \int_{-\infty}^{\infty} e^{\sigma x + \delta |x| + |h| |x|} dF(x) < \infty$$

in the interior of the strip of convergence. As $|h| \to 0$ the last expression tends to zero, so

$$\frac{\phi(\theta + h) - \phi(\theta)}{h} \to \int_{-\infty}^{\infty} x e^{\theta x} dF(x).$$

Thus $\phi'(\theta)$ exists for θ in the interior of the strip, which means that $\phi(\theta)$ is analytic there. □

Theorem 3.2. *The c.f. ϕ is uniformly continuous along vertical lines that belong to the strip of convergence.*

Proof. We have

$$|\phi(\sigma + i\omega_1) - \phi(\sigma + i\omega_2)| = \left| \int_{-\infty}^{\infty} e^{\sigma x} (e^{i\omega_1 x} - e^{i\omega_2 x}) dF(x) \right|$$

$$\leq \int_{-\infty}^{\infty} e^{\sigma x} |e^{i(\omega_1 - \omega_2)x} - 1| dF(x)$$

$$= 2 \int_{-\infty}^{\infty} e^{\sigma x} |\sin(\omega_1 - \omega_2)(x/2)| dF(x).$$

Since the integrand is uniformly bounded by $e^{\sigma x}$ and approaches 0 as $\omega_1 \to \omega_2$, uniformly continuity follows. □

Theorem 3.3. *An analytic c.f. is uniquely determined by its values on the imaginary axis.*

Proof. $\phi(i\omega)$ is the c.f. discussed in Chapter 2 and the result follows by the uniqueness theorem of that section. □

Theorem 3.4. *The function $\log \phi(\sigma)$ is convex in the interior of the strip of convergence.*

Proof.   We have

$$\frac{d^2}{d\sigma^2} \log \phi(\sigma) = \frac{\phi(\sigma)\phi''(\sigma) - \phi'(\sigma)^2}{\phi(\sigma)^2}$$

and by the Schwarz inequality

$$\phi'(\sigma)^2 = \left[\int_{-\infty}^{\infty} xe^{\sigma x}dF(x)\right]^2 = \left[\int_{-\infty}^{\infty} e^{\frac{1}{2}\sigma x} \cdot xe^{\frac{1}{2}\sigma x}dF(x)\right]^2$$

$$\leq \int_{-\infty}^{\infty} e^{\sigma x}dF(x) \cdot \int_{-\infty}^{\infty} x^2 e^{\sigma x}dF(x) = \phi(\sigma)\phi''(\sigma).$$

Therefore $\frac{d^2}{d\sigma^2} \log \phi(\sigma) \geq 0$, which shows that $\log \phi(\sigma)$ is convex.   □

Corollary 3.1. *If F has an analytic c.f. ϕ and $\phi'(0) = 0$, then $\phi(\sigma)$ is minimal at $\sigma = 0$. If ϕ is an entire function, then $\phi(\sigma) \to \infty$ as $\sigma \to \pm\infty$, unless F is degenerate.*

3.2. Moments

Recall that

$$\mu_n = \int_{-\infty}^{\infty} x^n dF(x), \quad \nu_n \int_{-\infty}^{\infty} |x|^n dF(x)$$

have been defined as the ordinary moment and absolute moment of order n respectively. If F has an analytic c.f. ϕ, then $\mu_n = \phi^{(n)}(0)$, and

$$\phi(\theta) = \sum_0^{\infty} \mu_n \frac{\theta^n}{n!},$$

the series being convergent in $|\theta| < \delta = \min(\alpha, \beta)$. The converse is stated in the following theorem.

Theorem 3.5. *If all moments of F exist and the series $\sum \mu_n \frac{\theta^n}{n!}$ has a nonzero radius of convergence ρ, then ϕ exists in $|\sigma| < \rho$, and inside the circle $|\theta| < \rho$,*

$$\phi(\theta) = \sum_0^{\infty} \mu_n \frac{\theta^n}{n!}.$$

Proof. We first consider the series $\sum \nu_n \frac{\theta^n}{n!}$ and show that it also converges in $|\theta| < \rho$. From Lyapunov's inequality

$$\nu_n^{\frac{1}{n}} \leq \nu_{n+1}^{\frac{1}{n+1}}$$

we obtain

$$\limsup \frac{\nu_n^{\frac{1}{n}}}{n} = \limsup \frac{\nu_{2n}^{\frac{1}{2n}}}{2n} = \limsup \frac{\mu_{2n}^{\frac{1}{2n}}}{2n} \leq \limsup \frac{|\mu_n|^{\frac{1}{n}}}{n}.$$

Also, since $|\mu_n| \leq \nu_n$ we have

$$\limsup \frac{|\mu_n|^{\frac{1}{n}}}{n} \leq \limsup \frac{\nu_n^{\frac{1}{n}}}{n}.$$

Therefore

$$\limsup \frac{|\mu_n|^{\frac{1}{n}}}{n} = \limsup \frac{\nu_n^{\frac{1}{n}}}{n}$$

which shows that the series $\sum \nu_n \frac{\theta^n}{n!}$ has radius of convergence ρ. For arbitrary $A > 0$ we have

$$\infty > \sum_0^\infty \nu_n \frac{|\theta|^n}{n!} \geq \sum^\infty \frac{|\theta|^n}{n!} \int_{-A}^A |x|^n dF(x) = \int_{-A}^A e^{|\theta x|} dF(x)$$

for $|\theta| < \rho$. So

$$\left| \int_{-A}^A e^{\theta x} dF(x) \right| \leq \int_{-A}^A e^{|\sigma x|} dF(x) < \infty$$

for $|\sigma| < \rho$. Since A is arbitrary, this implies that $\phi(\theta)$ converges in the strip $|\sigma| < \rho$. $\qquad\square$

3.3. The Moment Problem

The family of distributions given by

$$F_\varepsilon(x) = k \int_{-\infty}^x e^{-|y|^\alpha} \{1 + \varepsilon \sin(|y|^\alpha \tan \pi\alpha)\} dy$$

for $-1 \leq \varepsilon \leq 1$, $0 < \alpha < 1$ has the same moments of all orders. This raises the question: under what conditions is a distribution uniquely determined by its moments?

Theorem 3.6. *If F has an analytic c.f. then it is uniquely determined by its moments.*

Proof.   If F has an analytic c.f., then the series $\sum \mu_n \frac{\theta^n}{n}$ converges in $|\theta| < \rho = \min(\alpha, \beta)$ and $\phi(\theta)$ is given by this series there. If there is a second d.f. G with the same moments μ_n, then by Theorem 3.5, G has an analytic c.f. $\psi(\theta)$, and $\psi(\theta)$ is also given by that series in $|\theta| < \rho$. Therefore $\phi(\theta) = \psi(\theta)$ in the strip $|\sigma| < \rho$ and hence $F = G$.   □

The cumulant generating function

The principal value of $\log \phi(\theta)$ is called the cumulant generating function $K(\theta)$. It exists at least on the imaginary axis between $\omega = 0$ and the first zero of $\phi(i\omega)$. The cumulant of order r is defined by

$$K_r = i^{\frac{1}{r}} \left[\left(\frac{d}{d\omega} \right)^r \log \phi(i\omega) \right]_{\omega=0}.$$

This exists if, and only if, μ_r exists; K_r can be expressed in terms of μ_r. We have

$$K(i\omega) = \sum_0^\infty K_r \frac{(i\omega)^r}{r!}$$

whenever the series converges.

Theorem 3.7. *Let $\phi(\theta) = \phi_1(\theta)\phi_2(\theta)$, where $\phi(\theta)$, $\phi_1(\theta)$, $\phi_2(\theta)$ are c.f.'s. If $\phi(\theta)$ is analytic in $-\alpha < \sigma < \beta$, so are $\phi_1(\theta)$ and $\phi_2(\theta)$.*

Proof.   We have (with the obvious notations)

$$\int_{-\infty}^\infty e^{\sigma x} dF(x) = \int_{-\infty}^\infty e^{\sigma x} dF_1(x) \cdot \int_{-\infty}^\infty e^{\sigma x} dF_2(x),$$

and since $\phi(\sigma)$ is convergent, so are $\phi_1(\sigma)$ and $\phi_2(\sigma)$.   □

Theorem 3.8 (Cramér). *If X_1 and X_2 are independent r.v. such that their sum $X = X_1 + X_2$ has a normal distribution, then X_1, X_2 have normal distributions (including the degenerate case of the normal with zero variance).*

Proof. Assume without loss of generality that $E(X_1) = E(X_2) = 0$. Then $E(X) = 0$. Assume further that $E(X^2) = 1$. Let $\phi_1(\theta)$, $\phi_2(\theta)$ be the c.f.'s of X_1 and X_2. Then we have

$$\phi_1(\theta)\phi_2(\theta) = e^{\frac{1}{2}\theta^2}. \qquad (3.5)$$

Since the right side of (3.5) is an entire function without zeros, so are $\phi_1(\theta)$ and $\phi_2(\theta)$. By the convexity property (Theorem 3.4) we have $\phi_1(\sigma) \geq 1$, $\phi_2(\sigma) \geq 1$ as σ moves away from zero. Then (3.5) gives

$$e^{\frac{1}{2}\sigma^2} = \phi_1(\sigma)\phi_2(\sigma) \geq \phi_1(\sigma) \geq |\phi_1(\theta)|. \qquad (3.6)$$

Similarly $|\phi_2(\theta)| \leq e^{\frac{1}{2}\sigma^2}$. Therefore

$$e^{\frac{1}{2}\sigma^2}|\phi_1(\theta)| \geq |\phi_1(\theta)\phi_2(\theta)| = e^{\frac{1}{2}\mathrm{Re}(\theta^2)} = e^{\frac{1}{2}(\sigma^2 - \omega^2)},$$

so that

$$|\phi(\theta)| \geq e^{-\frac{1}{2}\omega^2}. \qquad (3.7)$$

From (3.6) and (3.7) we obtain

$$-\frac{1}{2}|\theta|^2 \leq -\frac{1}{2}\omega^2 \leq \log|\phi_1(\theta)| \leq \frac{1}{2}\sigma^2 \leq \frac{1}{2}|\theta|^2,$$

or, setting $K_1(\theta) = \log \phi_1(\theta)$,

$$|\mathrm{Re}\, K_1(\theta)| \leq \frac{1}{2}|\theta|^2. \qquad (3.8)$$

From a strengthened version of Liouville's theorem (see Lemma 3.1) it follows that $K_1(\theta) = a_1\theta + a_2\theta^2$. Similarly $K_2(\theta) = b_1\theta + b_2\theta^2$. □

Theorem 3.9 (Raikov). *If X_1 and X_2 are independent r.v. such that their sum $X = X_1 + X_2$ has a Poisson distribution, then X_1, X_2 have also Poisson distributions.*

Proof. The points of increase of X are $k = 0, 1, 2, \ldots$, so all points of increase α_1 and α_2 of X_1 and X_2 are such that $\alpha_1 + \alpha_2 = $ some k, and moreover the first points of increase of X_1 and X_2 are α and $-\alpha$ where α is some finite number. Without loss of generality we take

$\alpha = -\alpha = 0$, so that X_1 and X_2 have $k = 0, 1, 2, \ldots$ as the only possible points of increase. Their c.f.'s are then of the form

$$\phi_1(\theta) = \sum_0^\infty a_k e^{k\theta}, \quad \phi_2(\theta) = \sum_0^\infty b_k e^{k\theta} \qquad (3.9)$$

with a_0, $b_0 > 0$, a_k, $b_k \geq 0$ $(k \geq 1)$ and $\sum a_k = \sum b_k = 1$. Let $z = e^\theta$ and $\phi_1(\theta) = f_1(z)$, $\phi_2(\theta) = f_2(z)$. We have

$$f_1(z) f_2(z) = e^{\lambda(z-1)}. \qquad (3.10)$$

Therefore

$$a_0 b_k + a_1 b_{k-1} + \cdots + a_k b_0 = e^{-\lambda} \frac{\lambda^k}{k!} \quad (k = 0, 1, \ldots), \qquad (3.11)$$

which gives

$$a_k \leq \frac{1}{b_0} e^{-\lambda} \frac{\lambda^k}{k!}, \quad |f_1(z)| \leq \frac{1}{b_0} e^{\lambda(|z|-1)}. \qquad (3.12)$$

Similarly $|f_2(z)| \leq \frac{1}{a_0} e^{\lambda(|z|-1)}$. Hence

$$\frac{1}{a_0} e^{\lambda(|z|-1)} |f_1(z)| \geq |f_1(z) f_2(z)| = e^{\lambda(u-1)}$$

where $u = \operatorname{Re}(z)$. This gives

$$|f_1(z)| \geq a_0 e^{-\lambda(|z|-u)} \geq a_0 e^{-2\lambda|z|}. \qquad (3.13)$$

From (3.12) and (3.13), noting that $a_0 b_0 = e^{-\lambda}$ we find that

$$-2\lambda|z| \leq \log|f_1(z)| - \log a_0 \leq 2\lambda|z|,$$

or setting $K_1(z) = \log f_1(z)$, and $\log a_0 = -\lambda_1 < 0$,

$$|\operatorname{Re} K_1(z) + \lambda_1| \leq 2\lambda|z|. \qquad (3.14)$$

Proceeding as in the proof of Theorem 3.8, we obtain $\lambda_1 + K_1(z) = cz$, where c is a constant. Since $f_1 = 1$, $K_1(1) = 0$, so $c = \lambda_1$ and $f_1(z) = e^{\lambda_1(z-1)}$, which is the transform of the Poisson distribution. $\qquad \square$

Theorem 3.10 (Marcinkiewicz). *Suppose a distribution has a c.f. $\phi(\theta)$ such that $\phi(i\omega) = e^{P(i\omega)}$, where P is a polynomial. Then (i) $\phi(\theta) = e^{P(\theta)}$ in the whole plane, and (ii) ϕ is the c.f. of a normal distribution (so that $P(\theta) = \alpha\theta + \beta\theta^2$ with $\beta \geq 0$).*

Proof.   Part (i) is obvious. For Part (ii) let

$$P(\theta) = \sum_{1}^{n} a_k \theta^k, \quad n \text{ finite}, a_k \text{ real (cumulants)}.$$

From $|\phi(\theta)| \le \phi(\sigma)$ we obtain $|e^{P(\theta)}| \le e^{P(\sigma)}$ or $e^{\operatorname{Re} P(\theta)} \le e^{P(\sigma)}$. Therefore $\operatorname{Re} P(\theta) \le P(\sigma)$. Put $\theta = re^{i\alpha}$, so that $\sigma = r\cos\alpha$, $\omega = r\sin\alpha$. Then

$$a_n r^n \cos n\alpha + a_{n-1} r^{n-1} \cos(n-1)\alpha + \cdots$$
$$\le a_n r^n \cos^n \alpha + a_{n-1} r^{n-1} \cos^{(n-1)} \alpha + \cdots$$

Suppose $a_n \ne 0$. Dividing both sides of this inequality by r^n and letting $r \to \infty$ we obtain $a_n \cos n\alpha \le a_n \cos^n \alpha$. Putting $\alpha = \frac{\pi}{2n}$ we obtain

$$a_n \cdot 0 \le a_n \cos^n \frac{\pi}{2n},$$

so $a_n \ge 0$ for $n \ge 2$. Similarly, putting $\alpha = \frac{2\pi}{n}$ we find that

$$a_n \le a_n \cos^n \frac{2\pi}{n},$$

and since $\cos^n \frac{2\pi}{n} < 1$ for $n > 2$ we obtain $a_n \le 0$. Therefore $a_n = 0$ for $n > 2$, $P(\theta) = a_1 \theta + a_2 \theta^2$, and $\phi(\theta)$ is the c.f. of a normal distribution, the case $a_2 = 0$ being the degenerate case of zero variance. □

Theorem 3.11 (Bernstein). *Let X_1 and X_2 be independent r.v. with unit variances. Then if*

$$Y_1 = X_1 + X_2, \quad Y_2 = X_1 - X_2 \tag{3.15}$$

are independent, all four r.v. X_1, X_2, Y_1, Y_2 are normal.

This is a special case of the next theorem (with $n = 2$, $a_1 = b_1 = a_2 = 1$, $b_2 = -1$). For a more general result see [Feller (1971), pp. 77–80, 525–526]. He considers the linear transformation $Y_1 = a_{11}X_1 + a_{12}X_2$, $Y_2 = a_{21}X_1 + a_{22}X_2$ with $|\Delta| \ne 0$, where Δ is the determinant

$$\Delta = \begin{vmatrix} a_{11} & a_{12} \\ a_{21} & a_{22} \end{vmatrix}.$$

If $a_{11}a_{21} + a_{12}a_{22} = 0$ then the transformation represents a rotation. Thus (3.15) is a rotation.

Theorem 3.12 (Skitovic). *Let $X_1, X_2, \ldots, X_n$ be n independent r.v. such that the linear forms*

$$L_1 = a_1 X_1 + a_2 X_2 + \cdots + a_n X_n,$$
$$L_2 = b_1 X_1 + b_2 X_2 + \cdots + b_n X_n, \quad (a_i \neq 0, b_i \neq 0),$$

are independent. Then all the $(n+2)$ r.v. are normal.

Proof.   We shall first assume that (i) the ratios a_i/b_i are all distinct, and (ii) all moments of $X_1, X_2, \ldots, X_n$ exist. Then for α, β real we have (with obvious notations)

$$\phi_{\alpha L_1 + \beta L_2}^{(\theta)} = \phi_{\alpha L_1}^{(\theta)} \phi_{\beta L_2}^{(\theta)}$$

so that

$$\prod_{i=1}^{n} \phi_{(\alpha a_i + \beta b_i) X_i}^{(\theta)} = \prod_{i=1}^{n} \phi_{\alpha a_i X_i}^{(\theta)} \cdot \prod_{i=1}^{n} \phi_{\beta b_i X_i}^{(\theta)}.$$

Taking logarithms of both sides and expanding in powers of θ we obtain

$$\sum_{i=1}^{n} K_r^{(\alpha a_i + \beta b_i) X_i} = \sum_{i=1}^{n} K_r^{\alpha a_i X_i} + \sum_{i=1}^{n} K_r^{\beta b_i X_i}$$

or

$$\sum_{i=1}^{n} K_r^{(X_i)} \{ (\alpha a_i + \beta b_i)^r - (\alpha a_i)^r - (\beta b_i)^r \} = 0$$

for all $r \geq 1$. This can be written as

$$\sum_{i=1}^{n} K_r^{(X_i)} \sum_{s=1}^{r-1} \binom{r}{s} (\alpha a_i)^s (\beta b_i)^{r-s} = 0$$

for all $r \geq 1$ and all α, β. Hence

$$\sum_{i=1}^{n} a_i^s b_i^{r-s} K_r^{(X_i)} = 0 \quad (s = 1, 2, \ldots, r-1, r \geq 1),$$

Let $r \geq n+1$. Then for $s = 1, 2, \ldots, n$, $i = 1, 2, \ldots n$ we can write the above equations as

$$A_r \chi_r = 0 \tag{3.16}$$

where $A_r = \left(a_i^S b_i^{r-s}\right) 1 \leq s$, $i \leq n$ and χ_r is the column vector with elements $K_r^{(X_1)}, K_r^{(X_2)}, \ldots, K_r^{(X_n)}$. Since

$$|A_r| = (a_1 a_2 \cdots a_n)(b_1 b_2 \cdots b_n)^{r-1} \prod_{j>i}(c_j - c_i) \neq 0,$$

the only solution of (3.16) is $\chi_r = 0$. Therefore

$$K_r^{(X_i)} = 0 \quad \text{for } r \geq n+1, \quad i = 1, 2, \ldots, n. \tag{3.17}$$

Thus all cumulants of X_i of order $\geq n+1$ vanish, and $K^{(X_i)}(\theta)$ reduces to a polynomial of degree at most n. By the theorem of Marcinkiewicz, each X_i has a normal distribution. Hence L_1 and L_2 have normal distributions.

Next suppose that some of the a_i/b_i are the same. For example, let $a_1/b_1 = a_2/b_2$, and let $Y_1 = a_1 X_1 + a_2 X_2$. Then

$$L_1 = Y_1 + a_3 X_3 + \cdots + a_n X_n,$$
$$L_2 = \frac{b_1}{a_1} Y_1 + b_3 X_3 + \cdots + b_n X_n.$$

Repeat this process till all the a_i/b_i are distinct. Then by what has just proved, the Y_i are normal. By Cramér's theorem the X_i are normal.

Finally it remains to prove that the moments of X_i exist. This follows from the fact that L_1 and L_2 have finite moments of all orders. To prove this, we note that since $a_i \neq 0$, $b_i \neq 0$ we can take $a, c > 0$ such that $|a_i|, |b_i| \geq c > 0$. Also, let us standardize the a_i and b_i so that $|a_i| \leq 1, |b_i| \leq 1$. Now if $|L| = |a_1 X_1 + a_2 X_2 + \cdots + a_n X_n| \geq nM$, then at least one $|X_i| \geq M$. Therefore

$$P\{|L_1| \geq nM\} \leq \sum_{i=1}^{n} P\{|X_i| \geq M\}. \qquad (3.18)$$

Further, if $c|X_i| \geq nM$ and $|X_j| < M$ for all $j \neq i$, then $|L_1| \geq M$, $|L_2| \geq M$. Thus

$$P\{|L_1| \geq M, |L_2| \geq M\} \geq P\left\{|X_i| \geq \frac{nM}{c}\right\} \prod_{j \neq 1} P\{|X_j| < M\}$$

$$\geq P\left\{|X_i| \geq \frac{nM}{c}\right\} \prod_{j=1}^{n} P\{|X_j| < M\}.$$

Summing this over $i = 1, 2, \ldots, n$ we obtain, using (3.18),

$$n\,P\{|L_1| \geq M, |L_2| \geq M\} \geq P\left\{|L_1| \geq \frac{n^2 M}{c}\right\} \prod_{j=1}^{n} P\{|X_j| < M\}.$$

Since L_1 and L_2 are independent, this gives

$$\frac{P\left\{|L_1| \geq \frac{n^2 M}{c}\right\}}{P\{|L_1| \geq M\}} \leq n\frac{P\{|L_2| \geq M\}}{\prod_1^n P\{|X_j| < M\}} \to 0 \qquad (3.19)$$

as $M \to \infty$. We can write (3.19) as follows. Choose $n^2/c = \alpha > 1$. Then

$$\frac{P\{|L_1| \geq \alpha M\}}{P\{|L_1| \geq M\}} \to 0 \quad \text{as } M \to \infty. \qquad (3.20)$$

By a known result (Lemma 3.2), L_1, and similarly L_2, has finite moments of all orders.   □

Lemma 3.1 (see [Hille (1962)]). *If $f(\theta)$ is an entire function and $|\mathrm{Re} f(\theta)| \leq c|\theta|^2$, then $f(\theta) = a_1\theta + a_2\theta^2$.*

Proof.   We have $f(\theta) = \sum_0^\infty a_n\theta^n$, the series being convergent on the whole plane. Here

$$a_n = \frac{n!}{2\pi i} \int_{|\theta| \leq r} \frac{f(\theta)}{\theta^{n+1}} d\theta \quad (n = 0, 1, 2, \ldots). \qquad (3.21)$$

Also, since there are no negative powers

$$0 = \frac{n!}{2\pi i} \int_{|\theta| \leq r} f(\theta)\theta^{n-1}d\theta \quad (n = 1, 2, \ldots). \tag{3.22}$$

From (3.21) we obtain

$$a_n = \frac{n!}{2\pi i} \int_0^{2\pi} \frac{f(re^{i\alpha})}{r^{n+1}e^{i(n+1)\alpha}} re^{i\alpha}id\alpha$$

or

$$a_n r^n = \frac{n!}{2\pi} \int_0^{2\pi} f(re^{i\alpha})e^{-in\alpha}d\alpha \quad (n = 0, 1, \ldots). \tag{3.23}$$

Similarly from (3.22) we obtain

$$0 = \frac{n!}{2\pi} \int_0^{2\pi} f(re^{i\alpha})e^{in\alpha}d\alpha$$

or

$$0 = \frac{n!}{2\pi} \int_0^{2\pi} f(re^{i\alpha})e^{-in\alpha}d\alpha \quad (n = 1, 2, \ldots). \tag{3.24}$$

From (3.23) and (3.24) we obtain

$$a_n r^n = \frac{n!}{\pi} \int_0^{2\pi} \mathrm{Re} f(re^{i\alpha})e^{in\alpha}d\alpha \quad (n \geq 1).$$

Therefore

$$|a_n|r^n \leq \frac{n!}{\pi} \int_0^{2\pi} ck^2 d\alpha = 2cn!r^2$$

or

$$|a_n| \leq \frac{2cn!}{r^{n-2}} \to 0 \quad \text{as } r \to \infty \text{ for } n > 2.$$

This gives $f(\theta) = a_0\theta + a_1\theta + a_2\theta^2$. □

Lemma 3.2 (see [Loéve (1963)]). *For $\alpha > 1$ if*

$$\frac{1 - F(\alpha x) + F(-\alpha x)}{1 - F(x) + F(-x)} \to 0 \quad as \ x \to \infty$$

then F has moments of all orders.

Proof.   Given $\varepsilon > 0$ choose A so large that for $x > A$

$$\frac{1 - F(\alpha x) + F(-\alpha x)}{1 - F(x) + F(-x)} < \varepsilon \quad \text{and} \quad 1 - F(A) + F(-A) < \varepsilon.$$

Then for any positive integer r,

$$\frac{1 - F(\alpha^r A) + F(-\alpha^r A)}{1 - F(A) + F(-A)} = \prod_{s=1}^{r} \frac{1 - F(\alpha^s A) + F(-\alpha^s A)}{1 - F(\alpha^{s-1} A) + F(-\alpha^{s-1} A)} < \varepsilon^r$$

so that

$$1 - F(\alpha^r A) + F(-\alpha^r A) < \varepsilon^{r+1}.$$

Therefore

$$1 - F(x) + F(-x) < \varepsilon^{r+1}$$

for $x > \alpha^r A$. Now

$$\int_A^\infty n x^{n-1} [1 - F(x) + F(-x)] dx$$

$$= \sum_{r=0}^\infty \int_{\alpha^r A}^{\alpha^{r+1} A} n x^{n-1} [1 - F(x) + F(-\alpha)] dx$$

$$< \sum_{r=0}^\infty \varepsilon^{r+1} \int_{\alpha^r A}^{\alpha^{r+1} A} n x^{n-1} dx = \varepsilon A^n (\alpha^n - 1) \sum_0^\infty (\alpha^n \varepsilon)^r$$

and the series converges for $\varepsilon < \alpha^{-n}$. $\square$

3.4. Problems for Solution

1. If $1 - F(x) + F(-x) = 0(e^{-\rho x})$ as $x \to \infty$ for some $\rho > 0$, show
   that F is uniquely determined by its moments.
2. Show that the distribution whose density is given by

$$f(x) = \begin{cases} \dfrac{1}{2} e^{-|\sqrt{x}|} & \text{for } x > 0 \\ 0 & \text{for } x \leq 0 \end{cases}$$

   does not have an analytic c.f.

3. **Proof of Bernstein's theorem.** Introduce a change of scale so that $Y_1 = \frac{1}{\sqrt{2}}(X_1 + X_2)$, $Y_2 = \frac{1}{\sqrt{2}}(X_1 - X_2)$. Then prove that

$$K_s^{(Y_1)} = \left(\frac{1}{\sqrt{2}}\right)^s \left[1^s K_s^{(X_1)} + 1^s K_s^{(X_2)}\right],$$

$$K_s^{(Y_2)} = \left(\frac{1}{\sqrt{2}}\right)^s \left[1^s K_s^{(X_1)} + (-1)^s K_s^{(X_2)}\right],$$

and similarly for $K_s^{(X_1)}$, $K_s^{(X_2)}$ in terms of $K_s^{(Y_1)}$, $K_s^{(Y_2)}$. Hence show that

$$\left|K_s^{(X_i)}\right| \leq \frac{1}{2^s}\left\{2\left|K_s^{(X_1)}\right| + 2\left|K_s^{(X_2)}\right|\right\} \quad (i = 1, 2).$$

This gives $K_s^{(X_i)} = 0$ for $s > 2$, $i = 1, 2$.
4. If X_1, X_2 are independent and there exists one rotation $(X_1, X_2) \to (Y_1, Y_2)$ such that Y_1, Y_2 are also independent, then show that Y_1, Y_2 are independent for every rotation.

Chapter 4

Infinitely Divisible Distributions

4.1. Elementary Properties

A distribution and its c.f. ϕ are called infinitely divisible if for each positive integer n there exists a c.f. ϕ_n such that

$$\phi(\omega) = \phi_n(\omega)^n. \tag{4.1}$$

It is proved below (Corollary 4.1) that if ϕ is infinitely divisible, then $\phi(\omega) \neq 0$. Defining $\phi^{1/n}$ as the principal branch of the n-th root, we see that the above definition implies that $\phi^{1/n}$ is a c.f. for every $n \geq 1$.

Examples

(1) A distribution concentrated at a single point is infinitely divisible, since for it we have

$$\phi(\omega) = e^{ia\omega} = (e^{ia\omega/n})^n$$

where a is a real constant.

(2) The Cauchy density $f(x) = \frac{a}{\pi}[a^2 + (x - \gamma)^2]^{-1}$ $(a > 0)$ has $\phi(\omega) = e^{i\omega\gamma - a|\omega|}$. The relation (4.1) holds with $\phi_n(\omega) = e^{i\omega\gamma/n - a|\omega|/n}$. Therefore the Cauchy density is infinitely divisible.

(3) The normal density with mean m and variance σ^2 has c.f. $\phi(\omega) = e^{i\omega m - \frac{1}{2}\sigma^2\omega^2} = (e^{i\omega m/n - \frac{1}{2}\frac{\sigma^2}{n}\omega^2})^n$. Thus the normal distribution is infinitely divisible.

43

(4) The gamma distribution (including the exponential) is infinitely divisible, since its c.f. is

$$\phi(\omega) = (1 - i\omega/\lambda)^{-\alpha} = \left[(1 - i\omega/\lambda)^{-\alpha/n}\right]^n.$$

The discrete counterparts, the negative binomial and geometric distributions are also infinitely divisible.

(5) Let N be a random variable with the (simple) Poisson distribution $e^{-\lambda}\lambda^k/k!(k = 0, 1, 2, \ldots)$. Its c.f. is given by

$$\phi(\omega) = e^{\lambda(e^{i\omega}-1)},$$

which is clearly infinitely divisible. Now let $\{X_k\}$ be a sequence of independent random variables with a common c.f. Φ and let these be independent of N. Then the sum $X_1 + X_2 + \cdots + X_N - b$ has the c.f.

$$\phi(\omega) = e^{-i\omega b + \lambda[\Phi(\omega) - 1]},$$

which is the compound Poisson. Clearly, this is also infinitely divisible.

Lemma 4.1. Let $\{\phi_n\}$ be a sequence of c.f.'s. Then $\phi_n^n \to \phi$ continuous iff $n(\phi_n - 1) \to \psi$ with ψ continuous. In this case $\phi = e^\psi$.

Theorem 4.1. *A c.f. ϕ is infinitely divisible iff there exists a sequence $\{\phi_n\}$ of c.f.'s such that $\phi_n^n \to \phi$.*

Proof. If ϕ is infinitely divisible, then by definition there exists a c.f. ϕ_n such that $\phi_n^n = \phi(n \geq 1)$. Therefore the condition is necessary. Conversely, let $\phi_n^n \to \phi$. Then by Lemma 4.1, $n[\phi_n(\omega) - 1] \to \psi = \log\phi$. Now for $t > 0$,

$$e^{nt[\phi_n(\omega)-1]} \to e^{t\psi(\omega)} \text{ as } n \to \infty.$$

Here the expression on the left side is the c.f. of the compound Poisson distribution and the right side is a continuous function. Therefore for each $t > 0$, $e^{t\psi}$ is a c.f. and

$$\phi = e^\psi = (e^{\psi/n})^n,$$

which shows that ϕ is infinitely divisible. $\qquad\square$

Corollary 4.1. *If ϕ is infinitely divisible, $\phi \neq 0$.*

This was proved in the course of the proof of Theorem 4.1.

Corollary 4.2. *If ϕ is infinitely divisible, so is $\phi(\omega)^a$ for each $a > 0$.*

Proof.   We have $\phi^a = e^{a\psi} = (a^{a\psi/n})^n$. $\square$

Proof of Lemma 4.1. (i) Suppose $n(\phi_n - 1) \to \psi$ which is continuous. Then $\phi_n \to 1$ and the convergence is uniform in $\omega \in [-\Omega, \Omega]$. Therefore $|1 - \phi_n(\omega)| < \frac{1}{2}$ for $\omega \in [-\Omega, \Omega]$ and $n > N$. Thus $\log \phi_n$ exists for $\omega \in [-\Omega, \Omega]$, and $n > N$, and is continuous and bounded. Now

$$\log \phi_n = \log[1 + (\phi_n - 1)]$$
$$= (\phi_n - 1) - \frac{1}{2}(\phi_n - 1)^2 + \frac{1}{3}(\phi_n - 1)^3 - \cdots$$
$$= (\phi_n - 1)[1 + o(1)]$$

and therefore

$$n \log \phi_n = n(\phi_n - 1)[1 + o(1)] \to \psi$$

or $\phi_n^n \to \phi = e^{\psi}$.

(ii) Suppose $\phi_n^n \to \phi$. We shall first prove that ϕ has no zeros. It suffices to prove that $|\phi_n|^{2n} \to |\phi|^2$ implies $|\phi|^2 > 0$. Assume that this symmetrization has been carried out, so that $\phi_n^n \to \phi$ with $\phi_n \geq 0$, $\phi \geq 0$. Since ϕ is continuous with $\phi(0) = 1$, there exists an interval $[-\Omega, \Omega]$ in which ϕ does not vanish and therefore $\log \phi$ exists and is bounded. Therefore $\log \phi_n$ exists and is bounded for $\omega \in [-\Omega, \Omega]$ and $n > N$, so $n \log \phi_n \to \log \phi$. Thus $\log \phi_n \to 0$ or $\phi_n \to 1$. As in (i), $n(\phi_n - 1) \to \log \phi = \psi$. $\square$

Theorem 4.2. *If $\{\phi_n\}$ is a sequence of infinitely divisible c.f.'s and $\phi_n \to \phi$ which is continuous, then ϕ is an infinitely divisible c.f.*

Proof.   Since ϕ_n is infinitely divisible, $\phi_n^{1/n}$ is a c.f. Since

$$\left(\phi_n^{1/n}\right)^n \to \phi \text{ continuous,}$$

ϕ is an infinitely divisible c.f. by Theorem 4.1. $\square$

Theorem 4.3 (De Finetti). *A distribution is infinitely divisible iff it is the limit of compound Poisson distributions.*

Proof. If ϕ_n is the c.f. of a compound Poisson distribution, and $\phi_n \to \phi$ which is continuous, then by Theorem 4.2, ϕ is an infinitely divisible c.f. Conversely, let ϕ be an infinitely divisible c.f. Then by Theorem 4.1 there exists a sequence $\{\phi_n\}$ of c.f.'s such that $\phi_n^n \to \phi$. By Lemma 4.1

$$e^{n[\phi_n(\omega)-1]} \to e^\psi = \phi.$$

Here $e^{n[\phi_n(\omega)-1]}$ is the c.f. of a compound Poisson distribution. □

4.2. Feller Measures

A measure M is said to be a Feller measure if $M\{I\} < \infty$ for every finite interval I, and the integrals

$$M^+(x) = \int_{x_-}^\infty \frac{1}{y^2} M\{dy\}, \quad M^-(-x) = \int_{-\infty}^{-x_+} \frac{1}{y^2} M\{dy\} \quad (4.2)$$

converge for all $x > 0$.

Examples

(1) A finite measure M is a Feller measure, since

$$\int_{|y|>x} \frac{1}{y^2} M\{dy\} \leq \frac{1}{x^2}[M\{(-\infty,-x)\} + M\{(x,\infty)\}].$$

(2) The Lebesgue measure is a Feller measure, since

$$\int_{|y|>x} \frac{1}{y^2} dy = \frac{2}{x} \quad (x > 0).$$

(3) Let F be a distribution measure and $M\{dx\} = x^2 F\{dx\}$. Then M is a Feller measure with

$$M^+(x) = 1 - F(x_-), \quad M^-(-x) = F(-x_+).$$

Theorem 4.4. *Let M be a Feller measure, b a real constant and*

$$\psi(\omega) = i\omega b + \int_{-\infty}^\infty \frac{e^{i\omega x} - 1 - i\omega \sin x}{x^2} M\{dx\} \quad (4.3)$$

(the integral being convergent). Then corresponding to a given ψ there is only one measure M and one constant b.

Proof.   Consider

$$\psi^*(\omega) = \psi(\omega) - \frac{1}{2h}\int_{-h}^{h}\psi(\omega + s)ds \quad (h > 0). \tag{4.4}$$

We have

$$\psi^*(\omega) = \int_{-\infty}^{\infty}e^{i\omega x}\mu\{dx\} \tag{4.5}$$

where

$$\mu\{dx\} = \left(1 - \frac{\sin hx}{hx}\right)\frac{1}{x^2}M\{dx\} \tag{4.6}$$

and it is easily verified that μ is a finite measure. Therefore $\psi^*(\omega)$ determines μ uniquely, so M uniquely. Since $b = \operatorname{Im}\psi(1)$, the constant b is uniquely determined. $\qquad\square$

Convergence of Feller measures. Let $\{M_n\}$ be a sequence of Feller measures. We say that M_n converges properly to a Feller measure M if $M_n\{I\} \to M\{I\}$ for all finite intervals I of continuity of M, and

$$M_n^+(x) \to M^+(x), \quad M_n^-(-x) \to M^-(-x) \tag{4.7}$$

at all points x of continuity of M. In this case we write $M_n \to M$.

Examples

(1) Let $M_n\{dx\} = nx^2F_n\{dx\}$ where F_n is a distribution measure with weights $\frac{1}{2}$ at each of the points $\pm\frac{1}{\sqrt{n}}$. Then

$$M_n\{I\} = \int_I nx^2 F_n\{dx\} = n\left(\frac{1}{n}\cdot\frac{1}{2} + \frac{1}{n}\cdot\frac{1}{2}\right) = 1$$

if $\{-\frac{1}{\sqrt{n}}, \frac{1}{\sqrt{n}}\} \subset I$. Also $M_n^+(x) = M_n^-(-x) = 0$ for $x > \frac{1}{\sqrt{n}}$. Therefore $M_n \to M$ where M is a distribution measure concentrated at the origin. Clearly, M is a Feller measure.

(2) Let F_n be a distribution measure with Cauchy density $\frac{1}{\pi} \cdot \frac{n}{1+n^2 x^2}$ and consider $M_n\{dx\} = \pi n x^2 F_n\{dx\}$. We have

$$M_n\{(a,b)\} = \int_a^b \frac{n^2 x^2}{1 + n^2 x^2} \, dx \to |b - a|,$$

$$M_n^+(x) = \int_x^\infty \frac{n^2}{1 + n^2 y^2} \, dy \to \int_x^\infty \frac{dy}{y^2},$$

$$M_n^-(-x) = \int_{-\infty}^{-x} \frac{n^2}{1 + n^2 y^2} \, dy \to \int_{-\infty}^{-x} \frac{dy}{y^2}.$$

Therefore $M_n \to M$ where M is the Lebesgue measure.

Theorem 4.5. *Let $\{M_n\}$ be a sequence of Feller measures, $\{b_n\}$ a sequence of real constants and*

$$\psi_n(\omega) = i\omega b_n + \int_{-\infty}^\infty \frac{e^{i\omega x} - 1 - i\omega \sin x}{x^2} M_n\{dx\}. \qquad (4.8)$$

Then $\psi_n \to \psi$ continuous iff there exists a Feller measure M and a real constant b such that $M_n \to M$ and $b_n \to b$. In this case

$$\psi(\omega) = i\omega b + \int_{-\infty}^\infty \frac{e^{i\omega x} - 1 - i\omega \sin x}{x^2} M\{dx\}. \qquad (4.9)$$

Proof.   As suggested by (4.4)–(4.6) let

$$\mu_n\{dx\} = K(x) M_n\{dx\}, \quad \text{where } K(x) = x^{-2}\left(1 - \frac{\sin hx}{hx}\right) \qquad (4.10)$$

$$\mu_n = \mu_n\{(-\infty, \infty)\} < \infty. \qquad (4.11)$$

Then

$$M_n^*\{dx\} = \frac{1}{\mu_n} \mu_n\{dx\} \qquad (4.12)$$

is a distribution measure. We can write

$$\psi_n(\omega) = i\omega b_n + \mu_n \int_{-\infty}^\infty \frac{e^{i\omega x} - 1 - i\omega \sin x}{x^2} K(x)^{-1} M_n^*\{dx\}. \qquad (4.13)$$

(i) Let $M_n \to M$ and $b_n \to b$. Then

$$\mu_n \to \mu = \int_{-\infty}^{\infty} K(x) M\{dx\} > 0$$

and

$$M_n^* \to M^*, \quad \text{where } M^*\{dx\} = \frac{1}{\mu} K(x) M\{dx\}.$$

Therefore from (4.13) we find that

$$\psi_n(\omega) \to i\omega b + \mu \int_{-\infty}^{\infty} \frac{e^{i\omega x} - 1 - i\omega \sin x}{x^2} K(x)^{-1} M^*\{dx\}$$

$$= \psi(\omega).$$

(ii) Conversely, let $\psi_n(\omega) \to \psi(\omega)$ continuous. Then with $\psi_n^*(\omega)$, with $\psi(\omega)$ defined as in (4.4), $\psi_n^*(\omega) \to \psi^*(\omega)$; that is,

$$\int_{-\infty}^{\infty} e^{i\omega x} \mu_n\{dx\} \to \psi^*(\omega). \tag{4.14}$$

In particular

$$\mu_n = \mu_n\{(-\infty, \infty)\} \to \psi^*(0).$$

If $\psi^*(0) = 0$, then $\mu_n\{I\}$ and $M_n\{I\}$ tend to 0 for every finite interval I and by (i) $\psi(\omega) = i\omega b$ with $b = \lim b_n$. We have thus proved the required results in this case. Let $\mu = \psi^*(0) > 0$. Then (4.14) can be written as

$$\mu_n \int_{-\infty}^{\infty} e^{i\omega x} M_n^*\{dx\} \to \psi^*(\omega).$$

Therefore $M_n^* \to M^*$ where M^* is the distribution measure corresponding to the c.f. $\psi^*(\omega)/\psi^*(0)$. Thus

$$\mu_n \int_{-\infty}^{\infty} \frac{e^{i\omega x} - 1 - i\omega \sin x}{x^2} K(x)^{-1} M_n^*\{dx\}$$

$$\to \mu \int_{-\infty}^{\infty} \frac{e^{i\omega x} - 1 - i\omega \sin x}{x^2} K(x)^{-1} M^*\{dx\},$$

(the integrand being a bounded continuous function), and $b_n \to b$. Clearly,

$$M\{dx\} = \mu K(x)^{-1} M^*\{dx\}$$

is a Feller measure and

$$\psi(\omega) = i\omega b + \int_{-\infty}^{\infty} \frac{e^{i\omega x} - 1 - i\omega \sin x}{x^2} M\{dx\}$$

as required.   □

4.3. Characterization of Infinitely Divisible Distributions

Theorem 4.6. *A distribution is infinitely divisible iff its c.f. is of the form $\phi = e^{\psi}$, with*

$$\psi(\omega) = i\omega b + \int_{-\infty}^{\infty} \frac{e^{i\omega x} - 1 - i\omega \sin x}{x^2} M\{dx\}, \qquad (4.15)$$

M being a Feller measure, and b a real constant.

Proof.   (i) Let $\phi = e^{\psi}$ with ψ given by (4.15). We can write

$$\psi(\omega) = i\omega b - \frac{1}{2}\omega^2 M\{0\} + \lim_{\eta \to 0+} \psi_\eta(\omega) \qquad (4.16)$$

where

$$\psi_\eta(\omega) = \int_{|x|>\eta} \frac{e^{i\omega x} - 1 - i\omega \sin x}{x^2} M\{dx\}$$

$$= i\omega\beta + c\int_{|x|>\eta} (e^{i\omega x} - 1)G\{dx\}$$

with

$$cx^2 G\{dx\} = M\{dx\} \quad \text{for } |x| > \eta, \quad \text{and}$$

$$\beta = -\int_{|x|>\eta} \sin x \frac{M\{dx\}}{x^2},$$

c being determined so that G is a distribution measure. Let γ denote the c.f. of G; then

$$e^{\psi_\eta(\omega)} = e^{i\omega\beta + c[\gamma(\omega) - 1]}$$

is the c.f. of a compound Poisson distribution. As $\eta \to 0$, $\psi_\eta \to \psi_0$, where

$$\psi_0(\omega) = \int_{|x|>0} \frac{e^{i\omega x} - 1 - i\omega \sin x}{x^2} M\{dx\}$$

is clearly a continuous function. By Theorem 4.3, e^{ψ_0} is an infinitely divisible c.f. Now we can write

$$e^{\psi(\omega)} = e^{i\omega b - \frac{1}{2}\omega^2 M\{0\}} e^{\psi_0(\omega)},$$

so that ϕ is the product of $e^{\psi_0(\omega)}$ and the c.f. of a normal distribution. Therefore ϕ is infinitely divisible.

(ii) Conversely, let ϕ be an infinitely divisible c.f. Then by Theorem 4.3. ϕ is the limit of a sequence of compound Poisson c.f.'s. That is,

$$e^{c_n[\phi_n(\omega) - 1 - i\omega\beta_n]} \to \phi(\omega)$$

or

$$c_n \int_{-\infty}^{\infty} (e^{i\omega x} - 1 - i\omega\beta_n) F_n\{dx\} \to \log \phi(\omega)$$

where $c_n > 0, \beta_n$ is real and F_n is the distribution measure corresponding to the c.f. ϕ_n. We can write this as

$$\int_{-\infty}^{\infty} \frac{e^{i\omega x} - 1 - i\omega \sin x}{x^2} M_n\{dx\}$$

$$+ i\omega c_n \left(\int_{-\infty}^{\infty} \sin x F_n\{dx\} - \beta_n \right) \to \log \phi(\omega)$$

where $M_n\{dx\} = c_n x^2 F_n\{dx\}$. Clearly, M_n is a Feller measure. By Theorem 4.5 it follows that

$$M_n \to M \quad \text{and} \quad c_n \left(\int_{-\infty}^{\infty} \sin x F_n\{dx\} - \beta_n \right) \to b$$

where M is a Feller measure, b a real constant and

$$\log \phi(\omega) = i\omega b + \int_{-\infty}^{\infty} \frac{e^{i\omega x} - 1 - i\omega \sin x}{x^2} M\{dx\}.$$

This proves that $\phi = e^\psi$, with ψ given by (4.15). $\qquad\square$

Remarks.

(a) The centering function $\sin x$ is such that

$$\int_{-\infty}^{\infty} \frac{e^{ix} - 1 - i\sin x}{x^2} M\{dx\}$$

is real. Other possible centering functions are

$$\text{(i) } \tau(x) = \frac{x}{1+x^2}$$

and

$$\text{(ii) } \tau(x) = \begin{cases} -a & \text{for } x < -a, \\ |x| & \text{for } -a \le x \le a, \\ a & \text{for } x > a \quad \text{with } a > 0. \end{cases}$$

(b) The measure Λ (Lévy measure) is defined as follows: $\Lambda\{0\} = 0$ and $\Lambda\{dx\} = x^{-2}M\{dx\}$ for $x \ne 0$. We have

$$\int_{-\infty}^{\infty} \min(1, x^2)\Lambda\{dx\} < \infty,$$

as can be easily verified. The measure $K\{dx\} = (1+x^2)^{-1}M\{dx\}$ is seen to be a finite measure. This was used by Khintchine.

(c) The spectral function H is defined as follows:

$$H(x) = \begin{cases} -\displaystyle\int_{x-}^{\infty} \frac{M\{dy\}}{y^2} & \text{for } x > 0 \\ \displaystyle\int_{-\infty}^{x+} \frac{M\{dy\}}{y^2} & \text{for } x < 0, \end{cases}$$

H being undefined at $x = 0$. We can then write

$$\psi(\omega) = i\omega b - \frac{1}{2}w^2\sigma^2 + \int_{0+}^{\infty} [e^{i\omega x} - 1 - i\omega\tau(x)]dH(x)$$

$$+ \int_{-\infty}^{0-} [e^{i\omega x} - 1 - i\omega\tau(x)]dH(x),$$

where the centering function is usually $\tau(x) = x(1+x^2)^{-1}$. This is the so-called Lévy–Khintchine representation. Here H is non-decreasing in $(-\infty, 0)$ and $(0, \infty)$, with $H(-\infty) = 0$, $H(\infty) = 0$.

Also, for each $\varepsilon > 0$

$$\int_{|x|<\varepsilon} x^2\, dH(x) < \infty.$$

Theorem 4.7. *A distribution concentrated on $(0,\infty)$ is infinitely divisible iff its c.f. is given by $\phi = e^\psi$, with*

$$\psi(\omega) = i\omega b + \int_0^\infty \frac{e^{i\omega x} - 1}{x} P\{dx\} \qquad (4.17)$$

where $b \geq 0$ and P is a measure on $(0,\infty)$ such that $(1 + x)^{-1}$ is integrable with respect to P.

Theorem 4.8. *A function P is an infinitely divisible probability generation function (p.g.f.) iff $P(1) = 1$ and*

$$\log \frac{P(s)}{P(0)} = \sum_1^\infty a_k s^k \qquad (4.18)$$

where $a_k \geq 0$ and $\sum_1^\infty a_k = \lambda < \infty$.

Proof.   Let (4.18) hold and $P(1) = 1$. Put $a_k = \lambda f_k$ where $f_k \geq 0$, $\sum_1^\infty f_k = 1$. Let $F(s) = \sum_1^\infty f_k s^k$. Then

$$P(s) = P(0)e^{\lambda F(s)}$$

and

$$P(1) = P(0)e^\lambda.$$

Therefore

$$P(s) = e^{-\lambda + \lambda F(s)},$$

which is the p.g.f. of a compound Poisson distribution, and is therefore infinitely divisible.

Conversely, let P be an infinitely divisible p.g.f. Then Definition (4.1) implies $P(s)^{1/n}$ is also a p.g.f. (see [Feller (1968)]) for each $n \geq 1$. Let

$$P(s)^{1/n} = q_{0n} + q_{1n}s + q_{2n}s^2 + \cdots = Q_n(s) \quad \text{(say)}$$

or

$$P(s) = (q_{0n} + q_{1n}s + q_{2n}s^2 + \cdots)^n.$$

In particular $q_{0n}^n = P(0) = p_0$. If $p_0 = 0$, then $q_{0n} = 0$ and $P(s) = s^n(q_{1n} + q_{2n}s + q_{3n}s^2 + \cdots)^n$. This implies that $p_0 = p_1 = p_2 = \cdots = p_{n-1} = 0$ for each $n \geq 1$, which is absurd. Therefore $p_0 > 0$. It follows that $P(s) > 0$ and therefore $P(s)^{1/n} \to 1$ for $0 \leq s \leq 1$. Now

$$\frac{\log P(s) - \log P(0)}{-\log P(0)} = \frac{\log \sqrt[n]{\frac{P(s)}{P(0)}}}{\log \sqrt[n]{\frac{1}{P(0)}}} \sim \frac{\sqrt[n]{\frac{P(s)}{P(0)}} - 1}{\sqrt[n]{\frac{1}{P(0)}} - 1}$$

$$= \frac{\sqrt[n]{P(s)} - \sqrt[n]{P(0)}}{1 - \sqrt[n]{P(0)}} = \frac{Q_n(s) - Q_n(0)}{1 - Q_n(0)}.$$

Thus

$$\frac{Q_n(s) - Q_n(0)}{1 - Q_n(0)} \to \frac{\log P(s) - \log P(0)}{-\log P(0)}.$$

Here the left side is seen to be a p.g.f. By the continuity theorem the limit is the generating function of a non-negative sequence $\{f_j\}$. Thus

$$\frac{\log P(s) - \log P(0)}{-\log P(0)} = \sum_{1}^{\infty} f_j s^j = F(s) \quad \text{(say)}.$$

Putting $s = 1$ we find that $F(1) = 1$. Putting $\lambda = -\log P(0) > 0$ we obtain

$$P(s) = e^{-\lambda[1 - F(s)]}$$

which is equivalent to (4.18). $\qquad\square$

4.4. Special Cases of Infinitely Divisible Distributions

(A) Let the measure M be concentrated at the origin, with weight $\sigma^2 > 0$. Then (4.15) gives $\phi(\omega) = e^{i\omega b - \frac{1}{2}\omega^2\sigma^2}$, which is the c.f. of the normal distribution.

(B) Let M be concentrated at $h(\neq 0)$ with weight λh^2. Then

$$\phi(\omega) = e^{i\omega r + \lambda(e^{i\omega h} - 1)}, \quad r = b - \lambda \sin h.$$

Thus ϕ is the c.f. of the random variable $hN + r$, where N has the (simple) Poisson distribution $e^{-\lambda}\lambda^k/k!$ $(k = 0, 1, 2, \ldots)$.

(C) Let $M\{dx\} = \lambda x^2 G\{dx\}$ where G is the distribution measure with the c.f. ϕ. Clearly, M is a Feller measure and

$$\phi(\omega) = e^{i\omega\gamma + \lambda[\phi(\omega) - 1]}, \quad \gamma = b - \lambda \int_{-\infty}^{\infty} \sin x \ G\{dx\}.$$

We thus obtain the c.f. of a compound Poisson distribution.

(D) Let M be concentrated on $(0, \infty)$ with density $\alpha e^{-\lambda x} x (x > 0)$. It is easily verified that M is a Feller measure. We have

$$\int_0^\infty \frac{e^{i\omega x} - 1}{x^2} M\{dx\} = \alpha \int_0^\infty \frac{e^{-(\lambda - i\omega)x} - e^{-\lambda x}}{x}\{dx\}$$

$$= \alpha \log \frac{\lambda}{\lambda - i\omega} = \log\left(1 - \frac{i\omega}{\lambda}\right)^{-\alpha}.$$

Choosing

$$b = \alpha \int_0^\infty \frac{\sin x}{x} e^{-\lambda x}\, dx < \infty$$

we find that

$$\phi(\omega) = \left(1 - \frac{i\omega}{\lambda}\right)^{-\alpha},$$

This is the c.f. of the gamma density $e^{-\lambda x}\lambda^\alpha x^{\alpha - 1}/\Gamma(\alpha)$.

(E) Stable distributions. These are characterized by the measure M, where

$$M\{(-y, x)\} = C(px^{2-\alpha} + qy^{2-\alpha}) \quad (x > 0, y > 0)$$

where $C > 0$, $p \geq 0$, $q \geq 0$, $p + q = 1$, $0 < \alpha \leq 2$. If $\alpha = 2$, M is concentrated at the origin, and the distribution is the normal, as discussed in (A). Let $0 < \alpha < 2$, and denote by ψ_α the corresponding expression ψ. In evaluating it we choose an appropriate centering function $\tau_\alpha(x)$ depending on α. This changes the constant b and we obtain

$$\psi_\alpha(\omega) = i\omega\gamma + \int_{-\infty}^{\infty} \frac{e^{i\omega x} - 1 - i\omega\tau_\alpha(x)}{x^2} M\{dx\}$$

where

$$\gamma = b + \int_{-\infty}^{\infty} \frac{\tau_\alpha(x) - \sin x}{x^2} M\{dx\} \quad (|r| < \infty)$$

and

$$\tau_\alpha(x) = \begin{cases} \sin x & \text{if } \alpha = 1 \\ 0 & \text{if } 0 < \alpha < 1 \\ x & \text{if } 1 < \alpha < 2. \end{cases}$$

Substituting for M we find that

$$\psi_\alpha(\omega) = i\omega\gamma + c(2 - \alpha)[pI_\alpha(\omega) + q\bar{I}_\alpha(\omega)]$$

where

$$I_\alpha(\omega) = \int_0^{\infty} \frac{e^{i\omega x} - 1 - i\omega\tau_\alpha(x)}{x^{\alpha+1}} dx.$$

Evaluating the integral I_α we find that

$$\psi_\alpha(\omega) = i\omega\gamma - c|w|^\alpha \left[1 + i\beta\frac{\omega}{|\omega|}\Omega(|\omega|, \alpha)\right]$$

where $c > 0, |\beta| \le 1$ and

$$\Omega(|\omega|, \alpha) = \begin{cases} \tan\dfrac{\pi\alpha}{2} & \text{if } \alpha \ne 1 \\ \dfrac{2}{\pi}\log|w| & \text{if } \alpha = 1. \end{cases}$$

In Sec. 4.6 we shall discuss the detailed properties of stable distributions. We note that when $\beta = 0$ and $\alpha = 1$ we obtain $\psi_\alpha(\omega) = i\omega\gamma - c|\omega|$, so that ϕ is the c.f. of the Cauchy distribution.

4.5. Lévy Processes

We say a stochastic process $\{X(t), t \geq 0\}$ has *stationary independent increments* if it satisfies the following properties:

(i) For $0 \leq t_1 < t_2 < \cdots < t_n (n \geq 2)$ the random variables

$$X(t_1), X(t_2) - X(t_1), X(t_3) - X(t_2), \ldots, X(t_n) - X(t_{n-1})$$

are independent.

(ii) The distribution of the increment $X(t_p) - X(t_{p-1})$ depends only on the difference $t_p - t_{p-1}$.

For such a process we can take $X(0) \equiv 0$ without loss of generality. For if $X(0) \neq 0$, then the process $Y(t) = X(t) - X(0)$ has stationary independent increments, and $Y(0) = 0$.

If we write

$$X(t) = \sum_{k=1}^{n} \left[X\left(\frac{k}{n}t\right) - X\left(\frac{k-1}{n}t\right) \right] \tag{4.19}$$

then $X(t)$ is seen to be the sum of n independent random variables all of which are distributed as $X(t/n)$. Thus a process with stationary independent increments is the generalization to continuous time of sums of independent and identically distributed random variables.

A *Lévy process* is a process with stationary independent increments that satisfies the following additional conditions:

(iii) $X(t)$ is continuous in probability. That is, for each $\varepsilon > 0$

$$P\{|X(t)| > \varepsilon\} \to 0 \text{ as } t \to 0. \tag{4.20}$$

(iv) There exist left and right limits $X(t_-)$ and $X(t_+)$ and we assume that $X(t)$ is right-continuous: that is, $X(t_+) = X(t)$.

Theorem 4.9. *The c.f. of a Lévy process is given by $E[e^{i\omega X(t)}] = e^{t\psi(\omega)}$, where ψ is given by Theorem 4.6.*

Proof. Let $\phi_1(\omega) = E[e^{i\omega X(t)}]$. From (4.19) we find that $\phi_t(\omega) = [\phi_{t/n}(\omega)]^n$, so for each $t > 0$, ϕ_t is infinitely divisible and $\phi_t = e^{\psi_t}$. Also from the relation $X(t+s) \stackrel{d}{=} X(t) + X(s)$ we obtain the functional equation $\psi_{t+s} = \psi_t + \psi_s$. On account of (4.20), $\psi_t \to 0$ as $t \to 0$, so we must have $\psi_t(\omega) = t\psi_1(\omega)$. Thus $\phi_t(\omega) = e^{t\psi(\omega)}$ with $\psi = \psi_1$ in the required form.   □

Special cases: Each of the special cases of infinitely divisible distributions discussed in Sec. 4.4 leads to a Lévy process with c.f. $\phi_t(\omega) = e^{t\psi(\omega)}$ and ψ in the prescribed form. Thus for appropriate choices of the measure M we obtain the Brownian motion, simple and compound Poisson processes, gamma process and stable processes (including the Cauchy process).

A Lévy process with non-decreasing sample functions is called a *subordinator*. Thus the simple Poisson process and gamma process are subordinators.

4.6. Stable Distributions

A distribution and its c.f. are called stable if for every positive integer n there exist real numbers $c_n > 0$, d_n such that

$$\phi(\omega)^n = \phi(c_n\omega)e^{i\omega d_n}. \tag{4.21}$$

If $X, X_1, X_2, \ldots$ are independent random variables with the c.f. ϕ, then the above definition is equivalent to

$$X_1 + X_2 + \cdots + X_n \overset{d}{=} c_n X + d_n. \tag{4.22}$$

Examples

(A) If X has a distribution concentrated at a single point, then (4.22) is satisfied with $c_n = n$, $d_n = 0$. Thus a degenerate distribution is (trivially) stable. We shall ignore this from our consideration.

(B) If X has the Cauchy density $f(x) = \frac{a}{\pi}[a^2 + (x - r)^2]^{-1}$ $(a > 0)$, then $\phi(\omega) = e^{i\omega r - a|\omega|}$. The relation (4.21) holds with $c_n = n$, $dn = 0$. Thus the Cauchy distribution is stable.

(C) If X has a normal density with mean m and variance σ^2, then (22) holds with $c_n = \sqrt{n}$ and $d_n = m(n - c_n)$. Thus the normal distribution is stable.

The concept of stable distributions is due to Lévy (1924), who gave a second definition (see Problem 11).

Theorem 4.10. *Stable distributions are infinitely divisible.*

Proof.   The relation (4.21) can be written as

$$\phi(\omega) = \left[\phi\left(\frac{\omega}{c_n}\right) e^{-i\omega \frac{d_n}{n c_n}} \right]^n = \phi_n(\omega)^n$$

where ϕ_n is clearly a c.f. By definition ϕ is infinitely divisible.   □

Domains of attraction. Let $\{X_k, k \geq 1\}$ be a sequence of independent random variables with a common distribution F, and $S_n = X_1 + X_2 + \cdots + X_n$ ($n \geq 1$). We say that F belongs to the domain of attraction of a distribution G if there exist real constants $a_n > 0$, b_n such that the normed sum $(S_n - b_n)/a_n$ converges in distribution to G.

It is clear that a stable distribution G belongs to its own domain of attraction, with $a_n = c_n$, $b_n = d_n$. Conversely, we shall prove below that the only non-empty domains of attraction are those of stable distributions.

Theorem 4.11. *If the normed sum $(S_n - b_n)/a_n$ converges in distribution to a limit, then*

(i) *as $n \to \infty$, $a_n \to \infty$, $a_{n+1}/a_n \to 1$ and $(b_{n+1} - b_n)/a_n \to b$ with $|b| < \infty$, and*

(ii) *the limit distribution is stable.*

Proof.   (i) With the obvious notation we are given that

$$[\chi(\omega/a_n)e^{-i\omega b_n/na_n}]^n \to \phi(\omega) \tag{4.23}$$

uniformly in $\omega \in [-\Omega, \Omega]$. By Lemma 4.1 we conclude that

$$n[\chi(\omega/a_n)e^{-i\omega b_n/na_n} - 1] \to \psi(\omega)$$

where $\phi = e^{\psi}$. Therefore

$$\chi_n(\omega) = \chi(\omega/a_n)e^{-i\omega b_n/na_n} \to 1.$$

Let $\{a_{n_k}\}$ be a subsequence of $\{a_n\}$ such that $a_{n_k} \to a$ ($0 \leq a \leq \infty$).

If $0 < a < \infty$, then

$$1 = \lim |\chi(\omega/a_{n_k})| = |\chi(\omega/a)|,$$

while if $a = 0$, then

$$1 = \lim |\chi_{n_k} a_{n_k} \omega| = |\chi(\omega)|.$$

Both implications here would mean that χ is degenerate, which is not true. Hence $a = \infty$ and $a_n \to \infty$. From (4.23) we have

$$\chi \left(\frac{\omega}{a_{n+1}} \right)^{n+1} e^{-i\omega b_{n+1}/a_{n+1}} \to \phi(\omega),$$

which can be written as

$$\chi \left(\frac{\omega}{a_{n+1}} \right)^{n} e^{-i\omega b_{n+1}/a_{n+1}} \to \phi(\omega), \qquad (4.24)$$

since $\chi(\omega/a_{n+1}) \to 1$. By Theorem 2.10 it follows from (4.23) and (4.24) that $a_{n+1}/a_n \to 1$ and $(b_{n+1} - b_n)/a_n \to b$.

  (ii) For fixed $m \geq 1$ we have

$$\phi \left(\frac{\omega}{a_n} \right)^{mn} e^{-i\omega m b_n/a_n} = \left[\phi \left(\frac{\omega}{a_n} \right)^{n} e^{-i\omega b_n/a_n} \right]^{m} \to \phi^m(\omega).$$

Again by Theorem 2.10 it follows that $a_{mn}/a_n \to c_m$, $(b_{mn} - mb_n)/a_n \to d_m$, where $c_m > 0$ and d_m is real, while

$$\phi(\omega) = \phi^m \left(\frac{\omega}{c_m} \right) e^{-i\omega d_m/c_m}$$

or

$$\phi^m(\omega) = \phi(c_m \omega) e^{i\omega d_m}.$$

This shows that ϕ is stable. $\qquad \square$

Theorem 4.12. *A c.f. ϕ is stable iff $\phi = e^{\psi}$, with*

$$\psi(\omega) = i\omega\gamma - c|\omega|^{\alpha} \left[1 + i\beta \frac{\omega}{|\omega|} \Omega(|\omega|, \alpha) \right] \qquad (4.25)$$

where γ is real, $c > 0$, $0 < \alpha \leq 2$, $|\beta| \leq 1$ and

$$\Omega(|\omega|, \alpha) = \begin{cases} \tan \dfrac{\pi\alpha}{2} & \text{if } \alpha \neq 1 \\[2mm] \dfrac{2}{\pi}\log|\omega| & \text{if } \alpha = 1. \end{cases} \tag{4.26}$$

Here α is called the characteristic exponent of ϕ.

Proof. (i) Suppose ϕ is given by (4.25) and (4.26). Then for $a > 0$ we have

$$\begin{aligned} a\psi(\omega) - \psi(a^{1/\alpha}\omega) &= i\omega\gamma(a - a^{1/\alpha}) \\ &\quad -ac|\omega|^\alpha i\beta\frac{\omega}{|\omega|}[\Omega(|\omega|, \alpha) - \Omega(a^{1/\alpha}|\omega|, \alpha)] \\ &= \begin{cases} i\omega\gamma(a - a^{1/\alpha}) & \text{if } \alpha \neq 1 \\[2mm] i\omega\left(\dfrac{2\beta c}{\pi}\right)a\log a & \text{if } \alpha = 1. \end{cases} \end{aligned}$$

This shows that ϕ is stable.

(ii) Conversely, let ϕ be stable. Then by Theorem 4.11 it possesses a domain of attraction; that is, there exists a c.f. χ and real constants $a_n > 0$, b_n such that as $n \to \infty$

$$[\chi(\omega/a_n)e^{-i\omega b_n}]^n \to \phi(\omega).$$

Therefore by Lemma 4.1,

$$n[\chi(\omega/a_n)e^{-i\omega b_n} - 1] \to \psi(\omega)$$

where $\phi = e^\psi$. Let F be the distribution corresponding to χ. We first consider the case where F is symmetric; then $b_n = 0$. Let $M_n\{dx\} = nx^2 F\{a_n dx\}$. Then by Theorem 4.5 it follows that there exists a Feller measure M and a constant b such that

$$\psi(\omega) = i\omega b + \int_{-\infty}^{\infty} \frac{e^{i\omega x} - 1 - i\omega \sin x}{x^2} M\{dx\}. \tag{4.27}$$

Let

$$U(x) = \int_{-x}^{x} y^2 F\{dy\} \quad (x > 0). \tag{4.28}$$

Then

$$M_n\{(-x,x)\} = \frac{n}{a_n^2}U(a_nx) \to M\{(-x,x)\} \qquad (4.29a)$$

$$n[1 - F(a_nx)] = \int_{x_-}^{\infty} y^{-2}M_n\{dy\} \to M^+(x) \qquad (4.29b)$$

$$nF(-a_nx) = \int_{-\infty}^{-x_+} y^{-2}M_n\{dy\} \to M^-(-x). \qquad (4.29c)$$

By Theorem 4.11 we know that $a_n \to \infty$, $a_{n+1}/a_n \to 1$. Therefore $U(x)$ varies regularly at infinity and $M\{(-x,x)\} = Cx^{2-\alpha}$ where $C > 0$, $0 < \alpha \leq 2$. If $\alpha = 2$ the measure M is concentrated at the origin. If $0 < \alpha < 2$ the measure M is absolutely continuous.

In the case where F is unsymmetric we have

$$n[1 - F(a_nx + a_nb_n)] \to M^+(x), \quad nF(-a_nx + a_nb_n) \to M^-(x)$$

and an analogous modification of (4.29a). However it is easily seen that $b_n \to 0$, and so these results are fully equivalent to (4.29). Considering (4.29b) we see that either $M^+(x) \equiv 0$ or $1 - F(x)$ varies regularly at infinity and $M^+(x) = Ax^{-\alpha}$. Similarly $F(x)$ and $1 - F(x) + F(-x)$ vary regularly at infinity and the exponent α is the same for both M^+ and M^-. Clearly $0 < \alpha \leq 2$.

If M^+ and M^- vanish identically, then clearly M is concentrated at the origin. Conversely of M has an atom at the origin, then a symmetrization argument shows that M is concentrated at the origin, and M^+, M^- vanish identically. Accordingly, when $\alpha < 2$ the measure M is uniquely determined by its density, which is proportional to $|x|^{1-\alpha}$. For each interval $(-y,x)$ containing the origin we therefore obtain

$$M\{(-y,x)\} = C(px^{2-\alpha} + qy^{2-\alpha}) \qquad (4.30)$$

where $p + q = 1$. For $\alpha = 2$, M is concentrated at the origin. For $0 < \alpha < 2$ we have already shown in Sec. 4.4 that the measure (4.30) yields the required expression (4.25) for ψ. $\qquad\square$

Corollary 4.3. *If G_α is the stable distribution with the characteristic exponent α, then as $x \to \infty$*

$$x^\alpha[1 - G_\alpha(x)] \to Cp\frac{2-\alpha}{\alpha}, \quad x^\alpha G_\alpha(-x) \to Cq\frac{2-\alpha}{\alpha}. \quad (4.31)$$

Proof. Clearly, G_α belongs to its own domain of attraction with the norming constants $a_n = n^{1/\alpha}$. For $0 < \alpha < 2$, choosing $n^{1/\alpha}x = t$ in (4.29b) we find that $t^\alpha[1 - G_\alpha(t)] \to Cp\frac{2-\alpha}{\alpha}$ as $t \to \infty$. For $\alpha = 2$, G_α is the normal distribution and for it we have a stronger result, namely, $x^\beta[1 - G_\alpha(x)] \to 0$ as $x \to \infty$. □

Theorem 4.13. (i) *All stable distributions are absolutely continuous.*

(ii) *Let $0 < \alpha < 2$. Then moments of order $< \alpha$ exist, while moments of order $> \alpha$ do not.*

Proof. (i) We have $|\phi(\omega)| = e^{-c|\omega|^\alpha}$, with $c > 0$. Since the function is integrable over $(-\infty, \infty)$, the result (i) follows by Theorem 2.6(b).

(ii) For $t > 0$ an integration by parts gives

$$\int_{-t}^{t} |x|^\beta F\{dx\} = -t^\beta[1 - F(t) + F(-t)]$$

$$+ \int_0^t \beta x^{\beta-1}[1 - F(x) + F(-x)]dx$$

$$\le \int_0^t \beta x^{\beta-1}[1 - F(x) + F(-x)]dx.$$

If $\beta < \alpha$, this last integral converges as $t \to \infty$. Since by Corollary 4.3 we have $x^\alpha[1 - F(x) + F(-x)] \le M$ for $x > t$ where t is large. It follows that the absolute moment (and therefore the ordinary moment) of order $\beta < \alpha$ is finite. Conversely if the absolute moment of order $\beta > \alpha$ exists, then for $\varepsilon > 0$ we have

$$\varepsilon > \int_{|x|>t} |x|^\beta F\{dx\} > t^\beta[1 - F(t) + F(-t)]$$

or $t^\alpha[1 - F(t) + F(-t)] < \varepsilon t^{\alpha-\beta} \to 0$ as $t \to \infty$, which is a contradiction. Therefore absolute moments of order $\beta > \alpha$ do not exist. □

Remarks

(1) From the proof of Theorem 4.12 it is clear that

$$\phi(\omega)^a = \phi(c_a\omega)e^{i\omega d_a}$$

for all $a > 0$, and the functions c_a and d_a are given by

(i) $c_a = a^{1/\alpha}$ with $0 < \alpha \leq 2$, and

(ii) $d_a = \begin{cases} \gamma(a - a^{1/\alpha}) & \text{if } \alpha \neq 1 \\ (2\beta c/\pi)a \log a & \text{if } \alpha = 1. \end{cases}$

(2) If in the definition (4.21), $d_n = 0$, then the distribution is called *strictly stable*. However, the distinction between strict and weak stability matters only when $\alpha = 1$, because when $\alpha \neq 1$ we can take $d_n = 0$ without loss of generality. To prove this we note that $d_n = \gamma(n - n^{1/\alpha})$ for $\alpha \neq 1$, and consider the c.f.

$$\chi(\omega) = \phi(\omega)e^{-i\omega\gamma}.$$

We have

$$\chi(\omega)^n = \phi(\omega)^n e^{-i\omega n\gamma} = \phi(c_n\omega)e^{i\omega(d_n - n\gamma)}$$
$$= \chi(c_n\omega)e^{i\omega(c_n\gamma + d_n - n\gamma)} = \chi(c_n\omega)$$

which shows that χ is strictly stable.

(3) Let $\alpha \neq 1$ and assume that $\gamma = 0$. Then we can write

$$\psi(\omega) = -a|\omega|^\alpha \quad \text{for } \omega > 0, \text{ and } -\bar{a}|\omega|^\alpha \quad \text{for } \omega < 0 \quad (4.32)$$

where a is a complex constant. Choosing a scale so that $|a| = 1$ we can write $a = e^{i\frac{\pi}{2}\nu}$, where $\tan\frac{\pi\nu}{2} = \beta\tan\frac{\pi\alpha}{2}$. Since $|\beta| \leq 1$ it follows that

$$|\nu| \leq \alpha \quad \text{if } 0 < \alpha < 1, \quad \text{and} \quad |\nu| \leq 2 - \alpha \quad \text{if } 1 < \alpha < 2.$$
$$(4.33)$$

Theorem 4.14. *Let $\alpha \neq 1$ and let the c.f. of a stable distribution be expressed in the form*

$$\phi(\omega) = e^{-|\omega|^\alpha e^{\pm i\pi\nu/2}} \quad (4.34)$$

where in ± 1 the upper sign prevails for $\omega > 0$ and the lower sign for $\omega < 0$. Let the corresponding density be denoted by $f(x; \alpha, \nu)$. Then

$$f(-x; \alpha, \nu) = f(x; \alpha, -\nu) \quad \text{for } x > 0. \tag{4.35}$$

For $x > 0$ and $0 < \alpha < 1$,

$$f(x; \alpha, \nu) = \frac{1}{\pi x} \sum_{k=1}^{\infty} \frac{\Gamma(k\alpha + 1)}{k!} (-x^{-\alpha})^k \sin \frac{k\pi}{2}(\nu - \alpha) \tag{4.36}$$

and for $x > 0$ and $1 < \alpha < 2$

$$f(x; \alpha, \nu) = \frac{1}{\pi x} \sum_{k=1}^{\infty} \frac{\Gamma(k\alpha^{-1} + 1)}{k!} (-x)^k \sin \frac{k\pi}{2\alpha}(\nu - \alpha). \tag{4.37}$$

Corollary 4.4. *A stable distribution is concentrated on* $(0, \infty)$ *if* $0 < \alpha < 1$, $\nu = -\alpha$ *and on* $(-\infty, 0)$ *if* $0 < \alpha < 1$, $\nu = \alpha$.

Proofs are omitted.

Theorem 4.15. (a) *A distribution F belongs to the domain of attraction of the normal distribution iff*

$$U(x) = \int_{-x}^{x} y^2 F\{dy\} \tag{4.38}$$

varies slowly.

(b) *A distribution F belongs to the domain of attraction of a stable distribution with characteristic exponent $\alpha < 2$ iff*

$$1 - F(x) + F(-x) \sim x^{-\alpha} L(x) \quad (x \to \infty) \tag{4.39}$$

and

$$\frac{1 - F(x)}{1 - F(x) + F(-x)} \to p, \quad \frac{F(-x)}{1 - F(x) + F(-x)} \to q \tag{4.40}$$

where $p \geq 0$, $q \geq 0$ and $p + q = 1$, Here L is a slowly varying function on $(0, \infty)$; that is, for each $x > 0$

$$\frac{L(tx)}{L(t)} \to 1 \quad \text{as } t \to \infty. \tag{4.41}$$

The proof is omitted.

Theorem 4.16. *Let F be a proper distribution concentrated on $(0, \infty)$ and F_n the n-fold convolution of F with itself. If $F_n(a_nx) \rightarrow G(x)$, where G is a non-degenerate distribution, then $G = G_\alpha$, the stable distribution concentrated on $(0, \infty)$, with exponent $\alpha(0 < \alpha < 1)$, and moreover, $1 - F(t) \sim t^{-\alpha}L(t)/\Gamma(1-\alpha)$. Conversely, if $1 - F(t) \sim t^{-\alpha}L(t)/\Gamma(1-\alpha)$, we can find constants a_n such that $F_n(a_nx) \rightarrow G_\alpha(x)$. Here L is a slowly varying function.*

Proof.   (i) Suppose that $F_n(a_nx) \rightarrow G(x)$, and ϕ is the L.T. of G. Denote by F^* the L.T. of F. Then $F^*(\theta/a_n)^n \rightarrow \phi(\theta)$ or

$$-n \log F^*(\theta/a_n) \rightarrow -\log \phi(\theta).$$

This shows that $-\log F^*(\theta)$ is of regular variation at the origin, that is, $-\log F^*(\theta) \sim \theta^\alpha L(1/\theta)(\theta \rightarrow 0+)$, with $\alpha \geq 0$. Since $\log(1-z) \sim -z$ for small z, we find that $1 - F^*(\theta) \sim \theta^\alpha L(1/\theta)$. This gives $1 - F(t) \sim t^{-\alpha}L(t)/\Gamma(1-\alpha)$, as required. Moreover, $-\log \phi(\theta) = c\theta^\alpha(c > 0)$ or $\phi(\theta) = e^{-c\theta^\alpha}$, so that G is the stable distribution with exponent α. Here $0 < \alpha < 1$ since G is non-degenerate.

   (ii) Conversely, let $1 - F(t) \sim t^{-\alpha}L(t)/\Gamma(1-\alpha)(t \rightarrow \infty)$. This gives $1 - F^*(\theta) \sim \theta^\alpha L(1/\theta)(\theta \rightarrow 0+)$. Let us choose constants a_n so that $n[1 - F(a_n)] \rightarrow c/\Gamma(1-\alpha)$ for $0 < c < \infty$. Then as $n \rightarrow \infty$,

$$na_n^{-\alpha}L(a_n) = \frac{a_n^{-\alpha}L(a_n)}{[1 - F(a_n)]\Gamma(1-\alpha)} \cdot n[1 - F(a_n)]\Gamma(1-\alpha) \rightarrow c$$

and also

$$na_n^{-\alpha}L(a_n/\theta) = na_n^{-\alpha}L(a_n)\frac{L(a_n/\theta)}{L(a_n)} \rightarrow c.$$

Therefore $1 - F^*(\theta/a_n) \sim \theta^\alpha c/n$ and

$$F^*(\theta/a_n)^n = [1 - c\theta^\alpha/n + o(1/n)]^n \rightarrow e^{-c\theta^\alpha}.$$

This shows that $F_n(a_nx) \rightarrow G_\alpha(x)$. $\square$

4.7. Problems for Solution

1. Show that if F and G are infinitely divisible distributions so is their convolution $F * G$.

2. If ϕ is an infinitely divisible c.f., prove that $|\phi|$ is also an infinitely divisible c.f.

3. Show that the uniform distribution is not infinitely divisible. More generally, a distribution concentrated on a finite interval is not infinitely divisible, unless it is concentrated at a point.

4. Let $0 < r_j < 1$ and $\sum r_j < \infty$. Prove that for arbitrary a_j the infinite product

$$\phi(\omega) = \prod_{j=1}^{\infty} \frac{1 - r_j}{1 - r_j e^{i\omega a_j}}$$

converges, and represents an infinitely divisible c.f.

5. Let $X = \sum_1^{\infty} X_k/k$ where the random variables X_k are independent and have the common density $\frac{1}{2}e^{-|x|}$. Show that X is infinitely divisible, and find the associated Feller measure.

6. Let P be an infinitely divisible p.g.f. and ϕ the c.f. of an arbitrary distribution. Show that $P(\phi)$ is an infinitely divisible c.f.

7. If $0 \leq a < b < 1$ and ϕ is a c.f., then show that

$$\frac{1-b}{1-a} \cdot \frac{1-a\phi}{1-b\phi}$$

is an infinitely divisible c.f.

8. Prove that a probability distribution with a completely monotone density is infinitely divisible.

9. **Mixtures of exponential (geometric) distributions**. Let

$$f(x) = \sum_{k=1}^{n} p_k \lambda_k e^{-\lambda_k x}$$

where $p_k > 0$, $\sum p_k = 1$ and for definiteness $0 < \lambda_1 < \lambda_2 < \cdots < \lambda_n$. Show that the density $f(x)$ is infinitely divisible. (Similarly a mixture of geometric distributions is infinitely divisible.) By a limit argument prove that the density

$$f(x) = \int_0^{\infty} \lambda e^{-\lambda x} G(d\lambda),$$

where G is a distribution concentrated on $(0, \infty)$, is infinitely divisible.

10. If X, Y are two independent random variables such that $X > 0$ and Y has an exponential density, then prove that XY is infinitely divisible.

11. Show that a c.f. ϕ is stable if and only if given $c' > 0$, $c'' > 0$ there exist constants $c > 0$, d such that
$$\phi(c'\omega)\phi(c''\omega) = \phi(c\omega)e^{i\omega d}.$$

12. Let the c.f. ϕ be given by $\log \phi(\omega) = 2 \sum_{-\infty}^{\infty} 2^{-k}(\cos 2^k \omega - 1)$. Show that $\phi(\omega)^n = \phi(n\omega)$ for $n = 2, 4, 8, \ldots, \phi(\omega)$ is infinitely divisible, but not stable.

13. If $\phi(\omega)^2 = \phi(c\omega)$ and the variance is finite, show that $\phi(\omega)$ is stable (in fact normal).

14. If $\phi(\omega)^2 = \phi(a\omega)$ and $\phi(\omega)^3 = \phi(b\omega)$ with $a > 0$, $b > 0$, show that $\phi(\omega)$ is stable.

15. If F and G are stable with the same exponent α, so is their convolution $F * G$.

16. If X, Y are independent random variables such that X is stable with exponent α, while Y is positive and stable with exponent $\beta(< 1)$, show that $XY^{1/\alpha}$ is stable with exponent $\alpha\beta$.

17. **The Holtsmark distribution.** Suppose that n stars are distributed in the interval $(-n, n)$ on the real line, their locations $d_i(i = 1, 2, \ldots, n)$ being independent r.v. with a uniform density. Each star has mass unity, and the gravitational constant is also unity. The force which will be exerted on a unit mass at the origin (the gravitational field) is then
$$Y_n = \sum_{r=1}^{n} \frac{\text{sgn}(d_r)}{d_r^2}$$
Show that as $n \to \infty$, the distribution on Y_n converges to a stable distribution with exponent $\alpha = \frac{1}{2}$.

18. Let $\{X_k, k \geq 1\}$ be a sequence of independent random variables with the common density
$$f(x) = \begin{cases} \frac{2}{|x|^3} \log |x| & \text{for } |x| \geq 1 \\ 0 & \text{for } |x| \leq 1. \end{cases}$$
Show that $(X_1 + X_2 + \cdots + X_n)/\sqrt{n \log n}$ is asymptotically normal.

Chapter 5

Self-Decomposable Distributions; Triangular Arrays

5.1. Self-Decomposable Distributions

A distribution F and its c.f. ϕ are called self-decomposable if for every c in $(0, 1)$ there exists a c.f. ψ_c such that

$$\phi(\omega) = \phi(c\omega)\psi_c(\omega). \tag{5.1}$$

We shall call ψ_c the component of ϕ. Restriction of c to $(0, 1)$ is explained in Problem 5.1.

If ϕ is self-decomposable, then it can be proved that $\phi \neq 0$ (Problem 5.2). Thus the above definition implies that $\phi(\omega)/\phi(c\omega)$ is a c.f. for every c in $(0, 1)$.

Examples.

1. Degenerate distributions are (trivially) self-decomposable, and all their components are also degenerate.
2. A stable c.f. ϕ is self-decomposable, since by P. Lévy's second definition (Problem 4.11) we have

$$\phi(\omega) = \phi(c\omega)\phi(c'\omega)e^{i\omega d}$$

with $0 < c < 1, 0 < c' < 1$. Here $\psi_c(\omega) = \phi(c'\omega)e^{i\omega d}$ with c' and d depending on c; the component is also self-decomposable.

The concept of self-decomposable distributions is due to Khintchine (1936); they are also called distributions of class L.

Theorem 5.1. *If ϕ is self-decomposable, it is infinitely divisible, and so is its component ψ_c.*

Proof.   (i) Let $\{X_k, k \geq 1\}$ be independent random variables with X_k having c.f. $\psi_{k-1/k}(k\omega)$. Let $S_n = X_1 + X_2 + \cdots + X_n \ (n \geq 1)$. Then the c.f. of S_n/n is given by

$$E(e^{i\omega(S_n/n)}) = \prod_{k=1}^{n} \psi_{k-1/k}(k\omega/n)$$

$$= \prod_{k=1}^{n} \frac{\phi(k\omega/n)}{\phi((k-1)\omega/n)} = \phi(\omega)$$

so that ϕ is the c.f. of $X_1/n + X_2/n + \cdots + X_n/n$. By the theorem on triangular arrays ϕ is infinitely divisible.

(ii) We also have

$$\phi(\omega) = \phi(m\omega/n) \prod_{k=m+1}^{n} \psi_{k-1/k}(k\omega/n).$$

In this let $m \to \infty, n \to \infty$ in such a way that $m/n \to c \ (0 < c < 1)$. Then

$$\phi(\omega) = \phi(c\omega) \lim_{\substack{m \to \infty \\ n \to \infty}} \prod_{k=m+1}^{n} \psi_{k-1/k}(k\omega/n),$$

which shows that $\psi_c(\omega) = \lim_{m \to \infty} \prod_{k=m+1}^{n} \psi_{k-1/k}(k\omega/n)$. Again by the theorem on triangular arrays ψ_c is infinitely divisible. $\square$

As a converse of Theorem 5.1 we ask whether, given a sequence $\{X_k, k \geq 1\}$ of independent random variables there exist suitable constants $a_n > 0, b_n$ such that the normed sums $(S_n - b_n)/a_n$ converge in distribution. It is clear that in order to obtain this convergence we have to impose reasonable restrictions on X_k. We require that each component X_k/a_n become uniformly asymptotically negligible (**uan**) in the sense that given $\varepsilon > 0$ and $\Omega > 0$ one has for all sufficiently large n

$$|1 - E(e^{i\omega X_k/a_n})| < \varepsilon \quad \text{for } \omega \in [-\Omega, \Omega], k = 1, 2, \ldots, n \quad (5.2)$$

Theorem 5.2. *If the normed sums* $(S_n - b_n)/a_n$ *converge in distribution, then*

(i) *as* $n \to \infty, a_n \to \infty, a_{n+1}/a_n \to 1$; *and*
(ii) *the limit distribution is self-decomposable.*

Proof. (i) Let ϕ_k be the c.f. of X_k. We are given that

$$\chi_n(\omega) = e^{-i\omega b_n/a_n} \prod_{k=1}^{n} \phi_k(\omega/a_n) \to \phi(\omega) \qquad (5.3)$$

as $n \to \infty$. Take a subsequence $\{a_{n_k}\}$ of $\{a_n\}$ such that $a_{n_k} \to a(0 \le a \le \infty)$. If $0 < a < \infty$, then for each $k, |1 - \phi_k(\omega/a)| < \varepsilon$ for $\omega \in [-\Omega, \Omega]$ or

$$|1 - \phi_k(\omega)| < \varepsilon \quad \text{for } \omega \in \left[-\frac{\Omega}{a}, \frac{\Omega}{a} \right]$$

on account of the **uan** condition (5.2). Therefore $\phi_k(\omega) = 1$ in $[-\frac{\Omega}{a}, \frac{\Omega}{a}]$ and (5.3) gives $|\phi(\omega)| \equiv 1$. This means that ϕ is degenerate, which is not true. If $a = 0$, then

$$1 = |\phi(0)| = \lim_{k \to \infty} |\chi_{n_k}(a_{n_k}\omega)| = \lim \prod_{i}^{nk} |\phi_j(\omega)|.$$

This gives $|\phi_j(\omega)| = 1$ for all ω and again leads to degenerate ϕ. Therefore $a = \infty$, which means that $a_n \to \infty$. Proceeding as in the proof of Theorem 4.11 we find that $a_{n+1}/a_n \to 1$.

(ii) Given c in $(0, 1)$, for every integer n we can choose an integer $m < n$ such that $a_m/a_n \to c$, and $m \to \infty$, $n - m \to \infty$ as $n \to \infty$. We can write (5.3) as

$$\chi_n(\omega) = e^{-i\omega \frac{b_m}{a_n}} \prod_{k=1}^{m} \phi_k\left(\frac{a_m}{a_n} \cdot \frac{\omega}{a_m} \right) \cdot e^{-i\omega \frac{b_n - b_m}{a_n}}$$

$$\times \prod_{k=m+1}^{n} \phi_k\left(\frac{\omega}{a_n} \right)$$

$$= \chi_m\left(\frac{a_m}{a_n}\omega \right) \chi_{mn}(\omega) \quad \text{(say)}. \qquad (5.4)$$

Here $\chi_n(\omega) \to \phi(\omega)$ and $\chi_m((a_n/a_n)\omega) \to \phi(c\omega)$. If we prove that $\phi \neq 0$ then the c.f. $\chi_{mn}(\omega) \to \phi(\omega)/\phi(c\omega)$, a continuous function. It follows that $\phi(\omega)/\phi(c\omega)$ is a c.f., which means that ϕ is self-decomposable. To show that $\phi \neq 0$, note that $\phi(\omega_0) = 0$ for some ω_0 implies that $\phi(c\omega_0) = 0$. By induction $\phi(c^n\omega_0) = 0$, so $\phi(0) = 0$, which is absurd.   □

Theorem 5.3. *A c.f. is self-decomposable iff it is infinitely divisible and its Feller measure M is such that the two functions M_c^+, M_c^-, where*

$$M_c^+(x) = M^+(x) - M^+\left(\frac{x}{c}\right),$$

$$M_c^-(-x) = M^-(-x) - M^-\left(-\frac{x}{c}\right),$$

are monotone for every c in $(0,1)$.

5.2. Triangular Arrays

For each $n \geq 1$ let the random variables $X_{1n}, X_{2n}, \ldots, X_{r_n,n}$ be independent with X_{kn} having the distribution F_{kn} and c.f. ϕ_{kn}. The double sequence $\{X_{kn}, 1 \leq k \leq r_n, n \geq 1\}$, where $r_n \to \infty$, is called a *triangular array*. Let

$$S_{nn} = X_{1n} + X_{2n} + \cdots + X_{nn}.$$

We are interested in the limit distribution of $S_{nn} + \beta_n$ where $\{\beta_n\}$ is a sequence of real constants. The array $\{X_{kn}\}$ will be called a *null* array if it satisfies the uniformly asymptotically negligible (**uan**) condition: for each $\varepsilon > 0$ there exists a δ such that

$$P\{|X_{kn}| > \varepsilon\} < \delta \quad (k = 1, 2, \ldots, r_n) \tag{5.5}$$

for n sufficiently large. In terms of c.f.'s we can express this as follows: given $\varepsilon > 0$ and $\Omega > 0$ we have

$$|1 - \phi_{kn}(\omega)| < \varepsilon \quad \text{for } |\omega| < \Omega, \quad k = 1, 2, \ldots, r_n \tag{5.6}$$

for n sufficiently large. As special cases of the limit results we seek we have the following:

(i) Let $X_{kn} = \frac{X_k - b_n/n}{a_n} (k = 1, 2, \ldots, n; n \geq 1)$ where $\{X_k\}$ is a sequence of independent random variables with a common distribution. The problem is to find norming constants $a_n > 0, b_n$ such that the distribution of the random variables $(S_n - b_n)/a_n$ converges properly. We have seen that the limit distribution is (a) stable if the X_k are identically distributed, and (b) self-decomposable in general.

(ii) Let $P\{X_{kn} = 1\} = p_n$ and $P\{X_{kn} = 0\} = 1 - p_n$ and suppose that $p_n \to 0$, $np_n \to \lambda > 0$. Then we know that

$$P\{S_{nn} = k\} \to e^{-\lambda} \frac{\lambda^k}{k!} \quad (k = 0, 1, 2, \ldots).$$

We introduce the Feller measure M_n by setting

$$M_n\{dx\} = \sum_{k=1}^{r_n} x^2 F_{kn}\{dx\}. \tag{5.7}$$

For this we have

$$M_n^+(x) = \sum_{k=1}^{r_n} [1 - F_{kn}(x)], \quad M_n^-(-x) = \sum_{k=1}^{r_n} F_{kn}(-x) \tag{5.8}$$

for $x > 0$. We also introduce the truncation procedure by which a random variable X is replaced bu $\tau(X)$, where

$$\tau(x) = \begin{cases} -a & \text{for } x < -a, \\ x & \text{for } -a \leq x \leq a \\ a & \text{for } x > a. \end{cases} \tag{5.9}$$

It is seen that $E\tau(X + t)$ is a continuous monotone function of t and therefore vanishes for some t.

For each pair (k, n) there exists a constant t_{kn} such that $E\tau(X_{kn} + t_{kn}) = 0$. We can therefore center the random variable X_{kn} so that

$b_{kn} = E\tau(X_{kn}) = 0$. Assume this has been done, so that

$$b_{kn} = E\tau(X_{kn}) = 0. \tag{5.10}$$

Let

$$A_n = \sum_{k=1}^{r_n} E\tau^2(X_{kn}). \tag{5.11}$$

Proposition 5.1. *As* $n \to \infty$ *we have*

$$\log E\, e^{i\omega(S_{nn}+\beta_n)} = \sum_{k=1}^{r_n} [\phi_{kn}(\omega) - 1] + i\omega\beta_n + 0(A_n)$$

for $|\omega| < \Omega$.

Proof.   We have

$$\phi_{kn}(\omega) - 1 = \int_{-\infty}^{\infty} (e^{i\omega x} - 1) F_{kn}\{dx\}$$

$$= \int_{-\infty}^{\infty} [e^{i\omega x} - 1 - i\omega\tau(x)] F_{kn}\{dx\}$$

$$= \int_{-a}^{a} (e^{i\omega x} - 1 - i\omega x) F_{kn}\{dx\}$$

$$+ \int_{-\infty}^{-a} (e^{i\omega x} - 1 + i\omega a) F_{kn}\{dx\}$$

$$+ \int_{a}^{\infty} (e^{i\omega x} - 1 - i\omega a) F_{kn}\{dx\}$$

so that

$$|\phi_{kn}(\omega) - 1| \le \int_{-a}^{a} \frac{1}{2}\omega^2 x^2 F_{kn}\{dx\} + \int_{|x|>a} (2 + a|\omega|) F_{kn}\{dx\}$$

$$\le \frac{1}{2}\omega^2 E\tau^2(X_{kn}) + (2 + a|\omega|) E\tau^2(X_{kn}) a^{-1}$$

$$= c(\omega) E\tau^2(X_{kn}).$$

where $c(\omega) = \frac{1}{2}\omega^2 + \frac{2}{a} + |\omega|$. Summing this over $k = 1, 2, \ldots, r_n$ we obtain

$$\sum_{k=1}^{r_n} |\phi_{kn}(\omega) - 1| \le c(\omega) A_n. \tag{5.12}$$

The **uan** condition implies that $\phi_{kn}(\omega) \neq 0$ in $(-\Omega, \Omega)$ so that $\log \phi_{kn}$ exists in $(-\Omega, \Omega)$ for n sufficiently large. Therefore

$$\log \phi_{kn}(\omega) = \log[1 + \phi_{kn}(\omega) - 1]$$

$$= [\phi_{kn}(\omega) - 1] + \sum_{r=2}^{\infty} \frac{(-1)^{r-1}}{r} [\phi_{kn}(\omega) - 1]^r$$

and

$$\left| \sum_{k=1}^{r_n} \log \phi_{kn}(\omega) - \sum_{k=1}^{r_n} [\phi_{kn}(\omega) - 1] \right|$$

$$\leq \sum_{k=1}^{r_n} \sum_{r=2}^{\infty} \frac{1}{r} |\phi_{kn}(\omega) - 1|^r \leq \sum_{k=1}^{r_n} \frac{1}{2} \frac{|\phi_{kn}(\omega) - 1|^2}{1 - |\phi_{kn}(\omega) - 1|}$$

$$\leq \sup_{1 \leq k \leq r_n} |\phi_{kn}(\omega) - 1| \cdot \sum_{k=1}^{r_n} |\phi_{kn}(\omega) - 1|$$

$$< \varepsilon c(\omega) A_n \qquad\qquad\qquad (5.13)$$

by the **uan** condition and (5.2). From (5.2) and (5.3) it follows that

$$\log E \, e^{i\omega(S_{nn} + \beta_n)} = \log \left[e^{i\omega\beta_n} \prod_{k=1}^{r_n} \phi_{kn}(\omega) \right]$$

$$= \sum_{k=1}^{r_n} \log \phi_{kn}(\omega) + i\omega\beta_n$$

$$= \sum_{k=1}^{r_n} [\phi_{kn}(\omega) - 1] + i\omega\beta_n + 0(A_n)$$

as required.   □

Theorem 5.4. *Let $\{X_{kn}\}$ be a null array, centered so that $b_{kn} = 0$, and $\{\beta_n\}$ a sequence of real constants. Then $S_{nn} + \beta_n$ converges in distribution iff $\beta_n \to b$ and $M_n \to M$, a Feller measure. In this case the limit distribution is infinitely divisible, its c.f. ϕ being given by*

$\phi = e^\psi$, *with*

$$\psi(\omega) = i\omega b + \int_{-\infty}^{\infty} \frac{e^{i\omega x} - 1 - i\omega\tau(x)}{x^2} M\{dx\}. \qquad (5.14)$$

Proof. The desired result is a consequence of Theorem 4.5. In order to apply this theorem we define the distribution

$$F_n\{dx\} = \frac{1}{r_n} \sum_{k=1}^{r_n} F_{kn}\{dx\} \qquad (5.15)$$

and its c.f.

$$\phi_n(\omega) = \frac{1}{r_n} \sum_{k=1}^{r_n} \phi_{kn}(\omega). \qquad (5.16)$$

Then $M_n\{dx\} = r_n x^2 F_n\{dx\}$, the associated c.f. being $\phi_n(\omega) = e^{\psi_n(\omega)}$, where

$$\psi_n(\omega) = i\omega\beta_n + r_n[\phi_n(\omega) - 1]$$

$$= i\omega\beta_n + \sum_{k=1}^{r_n} [\phi_{kn}(\omega) - 1]. \qquad (5.17)$$

Using Proposition 5.1 we can therefore write

$$\log E\, e^{i\omega(S_{nn}+\beta_n)} = \psi_n(\omega) + 0(A_n) \quad (n \to \infty). \qquad (5.18)$$

(i) Let $M_n \to M$ and $\beta_n \to b$. By Theorem 4.5 it follows that $\psi_n \to \psi$ where

$$\psi(\omega) = i\omega b + \int_{-\infty}^{\infty} \frac{e^{i\omega x} - 1 - i\omega\tau(x)}{x^2} M\{dx\}. \qquad (5.19)$$

Furthermore

$$A_n = \sum_{k=1}^{r_n} E\tau^2(X_{kn}) = \sum_{k=1}^{r_n} \int_{-a}^{a} x^2 F_{kn}\{dx\} + \sum_{k=1}^{r_n} \int_{|x|\geq a} a^2 F_{kn}\{dx\}$$

$$= \int_{-a}^{a} M_n\{dx\} + \int_{|x|\geq a} a^2 \frac{1}{x^2} M_n\{dx\} \to M\{(-a, a)\}$$

$$+ [M^+(a) + M^-(-a)]a^2$$

if $(-a, a)$ is an interval of continuity of the measure M. Thus A_n tends to a finite limit and (5.8) gives $\log E\, e^{i\omega(S_{nn}+\beta_n)} \to \psi(\omega)$, with ψ given by (5.9).

(ii) Conversely, suppose that $S_{nn} + \beta_n$ converges in distribution. Then by (5.8) there is a c.f. ϕ such that

$$\psi_n(\omega) + 0(A_n) \to \log \phi(\omega) \tag{5.20}$$

for $|\omega| < \Omega$. Since the convergence is uniform we integrate (5.20) over $(-h, h)$ where $0 < h < \Omega$. The left side gives

$$\int_{-h}^{h} d\omega \sum_{k=1}^{r_n} (e^{i\omega x} - 1) F_{kn}\{dx\} + 2h0(A_n) + \beta_n \int_{-h}^{h} i\omega\, d\omega$$

$$= \sum_{k=1}^{r_n} \int_{-\infty}^{\infty} \left(\frac{2\sin hx}{x} - 2h \right) F_{kn}\{dx\} + 2h0(A_n)$$

and so

$$\sum_{k=1}^{r_n} \int_{-\infty}^{\infty} \left(1 - \frac{\sin hx}{hx} \right) F_{kn}\{dx\}$$

$$+ 0(A_n) \to -\frac{1}{2h} \int_{-h}^{h} \log \phi(\omega) \lambda\omega. \tag{5.21}$$

Now take $h < 2$; then

$$1 - \frac{\sin hx}{hx} \geq \frac{1}{10} h^2 x^2 \quad \text{for } |x| < 1,$$

$$1 - \frac{\sin hx}{hx} > \frac{1}{2} \quad \text{for } |x| \geq 1$$

Then the left side of (5.21) is

$$\geq \sum_{k=1}^{r_n} \frac{h^2}{10} \int_{-1}^{1} x^2 F_{kn}\{dx\} + \sum_{k=1}^{r_n} \frac{1}{2} \int_{|x|>1} F_{kn}\{dx\} + 0(A_n)$$

$$\geq \frac{h^2}{10} \sum_{k=1}^{r_n} \int_{-1}^{1} x^2 F_{kn}\{dx\} + \frac{1}{2} \sum_{k=1}^{r_n} \int_{|x|>1} F_{kn}\{dx\} + 0(A_n)$$

$$= \frac{h^2}{10} A_n + 0(A_n).$$

This shows that A_n is bounded as $n \to \infty$. We can therefore write (5.20) as

$$\psi_n(\omega) \to \log \phi(\omega) \tag{5.22}$$

uniformly in $|\omega| < \Omega$. The required result now follows from Theorem 4.5. $\square$

5.3. Problems for Solution

1. If $\phi(\omega) = \phi(c\omega)\psi_c(\omega)$ for $c \geq 1$, where $\psi_c(\omega)$ is a c.f., then either $\psi_c(\omega)$ is degenerate or $\phi(\omega)$ is degenerate.
   [If $c = 1$, then $\psi_c(\omega) \equiv 1$. If $c > 1$ then since $|\phi(\omega)| \leq |\phi(c\omega)|$ we obtain

   $$1 \geq |\phi(\omega)| \geq \left|\phi\left(\frac{\omega}{c}\right)\right| \geq \left|\phi\left(\frac{\omega}{c^2}\right)\right| \geq \cdots \geq \left|\phi\left(\frac{\omega}{c^n}\right)\right| \geq |\phi(0)| = 1,$$

   which gives $|\phi(\omega)| \equiv 1$.]

2. A self-decomposable c.f. ϕ never vanishes.
   [If $\phi(2a) = 0$ and $\phi(\omega) \neq 0$ for $0 \leq \omega < 2a$, then $\psi_c(2a) = 0$. We have

   $$|\psi_c(a)|^2 = |\psi_c(2a) - \psi_c(a)|^2 \leq 2[1 - \operatorname{Re}\psi_c(a)].$$

   Here $\psi_c(a) = \phi(a)/\phi(ca) \to 1$ as $c \to 1$, so we have a contradiction].

Bibliography

Feller, W., *An Introduction to Probability Theory and its Applications*, Vol. 1, 3rd Ed. (Wiley, New York, 1968).

Feller, W., *An Introduction to Probability Theory and its Applications*, Vol. 2, 2nd Ed. (Wiley, New York, 1971).

Hille, E., *Analytic Function Theory*, Vol. II (Boston, 1962).

Loéve, M., *Probability Theory*, 3rd Ed. (Van Nostrand, New York, 1963).

Index